Compass jellyfish.

Front cover upper: Tŵr Mawr Lighthouse, Ynys Llanddwyn, Anglesey © Alamy

Front cover lower: Common dolphin adult breaching for air in the Irish Sea off Pembrokeshire, Wales, UK © Drew Buckley.

Wonders of the Celtic Deep
Published in Great Britain in 2021 by Graffeg Limited.

Foreword & Introduction written by Dale Templar, Coasts: Life on the Edge & Shallows by Sally Weale, Into the Deep & Heroes by Anne Gallagher.

Graffeg Limited, 15 Neptune Court, Vanguard Way, Cardiff, CF24 5PJ, Wales, UK. Tel 01554 824000
www.graffeg.com

A CIP Catalogue record for this book is available from the British Library.

The publisher gratefully acknowledges the financial support of this book by the Books Council of Wales.
www.gwales.com

ISBN 9781802580075

1 2 3 4 5 6 7 8 9

The landmark BBC natural history series by One Tribe TV

Dale Templar, Sally Weale, Anne Gallagher

Wonders of the Celtic Deep

GRAFFEG

An offshore species, common dolphins are often sighted when coming in closer to the coast to feed.

Contents

“ ”

I feel so privileged to be a part of this life-changing series. As for the beauty and the wonder of it – it shows us things never before filmed and these film-makers have an eye for the wit and humour as well as the tragedy of life in the Deep.

Dame Siân Phillips
Actress

“ ”

This is British wildlife off the coast of Wales, shot in 4k, and it will totally blow you out of the water.

Dale Templar
Executive Producer and Managing Director of One Tribe TV

Dr Rohan Holt
Consultant and Safety Diver

“ ”

And it was three o'clock in the morning when we surfaced from successfully filming mussel beds. As we looked up, we had Comet Neowise just above us. It was a 'hair-standing-on-the-back-of-your-head' moment, and we were already euphoric about the dive. Seeing Neowise, on a beautiful, clear, calm, warm summer's night off the north Welsh coast, was just totally magic.

“ ”

I was amazed by the range of diversity in the Celtic Deep around Wales. I'm really proud that we did bring that sense of place to the series, showcasing the remarkable animals that live here.

Sally Weale
Series Producer

“ ”

We basically lived in Wales for a whole year, on location pretty much every day, just on the lookout for literally anything that we could film. We were just standing next to each other, just flying the drone, and we suddenly see this fin whale emerge from beneath the drone. As the whale breached, we were just speechless while everyone else on the boat started cheering. That was the shot where the whale comes up out of the water and you can just see its whole mouth open to catch the fish. That moment was pretty special.

Anne Gallagher
Series Director

About the Series

An unlikely resident of Welsh waters, the basking shark is the second largest fish in the world.

Back at the start of 2019, life seemed pretty normal in Wales. That summer, holiday-makers from Wales and beyond flocked to the coast to enjoy the delights of the country's magnificent ocean habitat. Wales is, of course, very much a coastal nation, surrounded on three sides by water, with over 2,700 km of coastline (including Anglesey but excluding the islands).

During the year, Welsh beaches won many accolades for cleanliness and high environmental standards and Tenby was named UK Beach of the Year by *The Sunday Times*. It got me thinking, was this a perfect time to make a natural history programme that showcased the animals and habitats that few tourists ever get to see or enjoy? Was this the right time to make *Wonders of the Celtic Deep*?

I was spurred on by the airing of the first ever landmark natural history documentary series made for BBC Cymru Wales called *Land of the Wild* (Plimsoll Productions), which set out to show off the country in a way that had never been done before, primarily celebrating its land-based wildlife.

I knew very well how stunning and varied the coastline is, but was there enough to offer the BBC a new four-part natural history series about our wonderful and often hidden water-world. The research started and very soon we began to pull together a remarkable wildlife 'wish list' for the series that has become *Wonders of the Celtic Deep*. We also discovered that Wales has as many underwater water habitats as there are topside landscapes, each attracting a vast number of different species. Our Welsh waters seemed to have everything we needed to fill a four-hour series for primetime viewing. I wanted to characterise the habitats in the same way we normally do with land-based environments; sandy beds became the deserts; kelp forests became the rainforests; seagrass obviously became grasslands; rocks became mountains and shipwrecks became submarine cities. So little has ever been filmed for broadcast in British, let alone Welsh waters, that the animals practically screamed out to be filmed; from some of the smallest and most quirky animals on earth, like the colourful sex-changing cuckoo wrasse, to the second largest fish on earth, the iconic basking shark, up to the absolutely gigantic fin whale, growing up to 20 m in length and second only in size to the blue whale. In the early days the series was actually called 'Blue Wales', but whenever we got on the phone to discuss the series people thought we were making a film about them – so that pun soon got dropped!

So we knew we had a great idea, but could we pull it off?

So we knew we had a great idea, but could we pull it off? The first major hurdle is financing. Making what's called 'blue chip' natural history means focusing in on animals, filming extraordinary behaviour to build great stories about them, shot with top-end cameras and kit. I was lucky enough to have worked at the BBC Natural History Unit (NHU) for a decade and knew all about the costs involved in making landmark natural history. Back in the noughties I watched my colleagues making *Planet Earth* and *Blue Planet*, then I myself spent four years running *Human Planet* at much the same time as *Frozen Planet* was being filmed. I also knew the very specific challenges involved when trying to capture marine-based wildlife. *Human Planet* filmed animals and humans in and around water in several episodes, including the arctic, rivers and, of course, the oceans episode, 'Into the Blue'. Just before starting *Human Planet*, I had made a high-budget expedition series for the NHU called *Pacific Abyss*, which took me several thousand kilometres across the ocean on a large boat with some of the very best specialist dive camera operators in the world. In short, I thought I knew the many risks, costs and frustrations that would inevitably come when trying to film *Wonders of the Celtic Deep*. Filming animals on land is one thing, filming animals underwater is another. After the success of *Land of the Wild*, the bar was high, but BBC Wales loved the scale and ambition of *Wonders of the Celtic Deep* and commissioned the series. I found additional funding from the brilliantly supportive Creative Wales and international distributor Orange Smarty also came in. Even so, I still wasn't in the financial 'comfort zone' I was used to when working at the NHU.

However, I was now absolutely passionate and determined to deliver this series. For the first time, I wanted to show people in Wales just how spectacular the marine and other ocean-dependent life is on their own doorstep and to share this with everyone across the UK and the rest of the world. Film-makers and holiday-makers spend a fortune going to the other side of the planet to see iconic marine animals, but it's just as easy to see them at home. I have been fortunate enough to dive along the Great Barrier Reef several times and the inner reefs are particularly disappointing and often lacking in fish and colourful corals. The Sea of Cortez may be called the aquarium of the world, but you can spot plenty of large whales and swim with sharks in Wales. The tip of South Africa may have the sardine run, but there are bait ball spectacles in the Irish Sea all summer. Right now, I can hear readers shouting at me: ' but what about the weather?' Indeed, the weather is always challenging in Wales, changing in a heartbeat. I put together a team that included experts at reading Welsh weather and predicting its impact on the Welsh waters. They were extremely flexible and opportunistic, often, like the fish we were filming, prepared to work day or night, seven days a week.

The production finally started in January 2020 and I had a great team in place who I had somehow persuaded to take on this almost impossible challenge on a budget with absolutely no buffers in it. At the helm, I took on the role of executive producer, with Sally Weale as series producer.

Above: Director of Photography Rob Taylor during underwater filming.

We had less than a year to film, and so much we wanted to film was seasonal, only happening once in very small filming windows.

Sally is a brilliant natural historian and TV storyteller who had worked with me at the BBC. We took on two young marine biologists, one from Bangor University, to do the crucial research. They were supported by members of One Tribe TV's core team, based in Cardiff, including assistant producer Cameron Howells and production co-ordinator Francesca Barbieri.

Technically, we had decided to jump in the metaphorical deep end by offering BBC Wales its first ever major 4K factual production. 4K comes with a whole range of extra production costs; cameras, lenses, storage of data, but the quality is superb and I wanted the series to be future-proofed. I wanted a dedicated Director of Photography and had already approached Rob Taylor, an adventure and underwater natural history cameraman. Rob and I were good friends, having worked together on a number of series, including *Extreme Wales with Richard Parks* (BBC Wales) and *Bear Grylls: Mission Survive* (ITV1); I knew Rob would be up for the challenge and wouldn't be put off by some cold water and rough seas. I was right, and he soon proved himself to be *Celtic Deep*'s answer to the Man from Atlantis.

The production would inevitably throw up complicated logistics and complex health and safety considerations, so also needed a first-class location manager. Here, I called on another person I'd worked with before and trusted, Jethro (Jet) Moore. Jet runs Adventure Beyond, based in Ceredigion, and has more water qualifications and safety knowledge than anyone I know, and an intimate working knowledge of the Welsh coast. In addition, Jet has more enthusiasm and 'can do' attitude than most human beings on the planet.

Many other human superstars joined the team in our efforts to film animal superstars in action – how could anything go wrong? Then along came a global pandemic. Lockdown.

The impact of Covid on the production cannot be underestimated. Although TV production was classed as an essential service, it was like trying to film a wildlife series with one fin tied behind your back. On the plus side, we were mainly filming animals and not humans – animals can't catch Covid. On the minus side Covid issues caused filming delays and cancellations. Of course, the animals carried on doing their behavioural stuff regardless; lockdown or no lockdown they weren't waiting for us. We had less than a year to film, and so much we wanted to film was seasonal, only happening once in very small filming windows. The first things to go were the islands. Though we had fully advanced plans to film on Skokholm, Skomer and Bardsey, one by one they fell by the wayside, along with the stories we wanted to capture there. Then the local authorities in each area of Wales set out different rules and regulations about what we could and couldn't do.

At one time, when the ports started to close down, we couldn't get any kind of motorised boat out to sea, and these rules kept changing from one day to the next. Frustratingly, Fran and the team spent far more time sorting out permissions to film than filming. Stories changed as we jumped on every opportunity that opened up, especially when restrictions started to relax over the summer and autumn.

We obviously had to put in place our own strict Covid filming guidelines and decided that the best thing we could do was create a filming and production bubble, as they started to do in drama productions. Jet very kindly offered us use of an amazing, if slightly dilapidated, farmhouse on the stunning coastal cliffs near Cardigan. For several months last summer this is where most of the *Celtic Deep* team both lived and worked, surrounded by tonnes of expensive filming kit. It came complete with baby cow who had been abandoned by its mum.

Director of Photography Rob Taylor didn't realise his experience of working for years in remote locations would come into play in Wales: 'Trying to make things happen on a small budget was difficult, but Covid raised things to the next level. We were having to do things in quite an unusual way. We all lived in a *Withnail and I* farmhouse which, in the nicest possible way, was only habitable if you'd spent time living in the Peruvian jungle! Just living there was kind of ridiculous, but added to that was the ridiculousness of having to look after a cow. Dale, the baby cow, thought she was a person. She used to see us going in the house, so she wanted to come in. One morning I was loading the car up with kit and I left the door open. Next thing I knew I found her in the kitchen.'

Dale the cow was indeed named after me, while still a youngster. She was now the size of a Great Dane, incredibly headstrong and rather feisty: 'Safety diver Nicki and I had to push her through the house and out of the back door. It took ages, she was kicking and biting. She even tried to go upstairs at one point. So we ended up being late on location for one of our very first shoots because we were having to wrangle a cow!'

Once the team did start on location, flexibility and thinking 'out of the box' was key to our success.

Very often we went to film one animal or behaviour and found something completely different, but with so few filming opportunites we learnt to go with the flow. Whatever the Celtic Deep offered up we filmed, and the results were often incredible.

Sometimes, even when we knew exactly where the animals were, they weren't always easy to get to. Rob and Jet are both extremely experienced paddlers and used kayaks, canoes and inflatables to get to places that couldn't be reached even by small boats. Without noisy engines, they were often able to get much closer to animals. Jet used canoes in a very early shoot to reach moulting seals on a small beach under an inaccessible cliff and remembers the moment his heart stopped when they tried to get off the beach at the end of a long filming day:

'We had rafted canoes, which are very, very stable, the chance of them tipping is pretty slim to zero, but they can swamp, as we found out, when a freak wave hit us and filled the canoe with probably £80,000 worth of our filming equipment in it.'

Fortunately, every piece of kit had been packed away in strong waterproof Peli cases. 'You can imagine the relief when we got back to the farmhouse and opened each box one by one. Luckily enough, other than a few drips that got through the seals, everything was okay.'

While scuba diving allowed the team to film deep underwater for periods of up to an hour or more, sometimes Rob wanted to film while free-diving, without the speed and manoeuvring constraints of scuba gear. By holding his breath, Rob was able to capture some of the most magical footage in the series, the sequence with the flirtatious grey seals in episode one being a prime example. It was filmed off the Smalls Lighthouse, which sits on a clump of rocks some 32 km off the coast of Pembrokeshire. He'd go underwater for a couple of minutes at a time and also film from the surface. The seals were so friendly, playful, gentle and curious that after filming he told me he felt as if he was part of a *ménage à trois*!

At the same time, Jet was with a second camera operator, Rich Stevenson: 'The seals we were filming were getting quite playful and starting to rub against us. And then this female came along and she took a nibble in a certain spot in my wetsuit, very close to a certain important part of my anatomy.' After that, Jet and Rich swam back to the boat as quickly as possible, but not before the female took another playful love bite. Only the wetsuit and Jet's ego were damaged.

Without any doubt the most heart-stopping and thrilling filming that took place in the series was the bait ball sequence in the 'Deep' episode. We had to film whales in Wales but we all knew this was high risk. The budget could only afford a few days of a full dive crew living out on a boat in the Celtic Deep. With us was Nicki Meharg, acting as a safety diver; she is an expert in cetaceans and runs trips from her base in Pembrokeshire. While bait ball feeding frenzies happen under the water, to find them you have to start by looking up in the sky. Sea birds, particularly gannets, attack the fish using an aerial bombardment. Just before 6am on the first morning, the watch team spotted some bird action, and as the boat got closer they could see whales. By now the whole team was awake and ready to roll. To get close to the bait ball of small fish that was under attack, they needed to move onto a dingy for maximum manoeuvrability and minimum disturbance.

Rob Taylor will never forget that morning:

'Anne [series director] was driving our little dingy in the open ocean. In front

Without any doubt the most heart-stopping and thrilling filming that took place in the series was the bait ball sequence, in the 'Deep' episode.

of us were the birds and whales. Then suddenly, very close-by, this giant long fin came out of the water. Anne, Nicki and I thought that's not a fin of a fin whale or a minke or any dolphin or shark or anything like that. We just didn't know what has a fin that big, other than a killer whale. Then a few minutes later it came through again. That's an orca fin, was the only thing we could think of. And then we looked at each other and said if there's an orca, then we know it's on a bait ball of fish. So, we got in and swam towards this bait ball. Here in the open ocean, you realise you're just tiny. There were bits of dead fish everywhere, glistening in the sun, and guillemots swimming past us. Then I feel this big push off my back – I literally thought a killer whale was attacking me. I looked around, expecting the worst.' That morning, the team were swimming closer together than usual and Rob had actually been attacked by Jet!

It turned out that they hadn't been joined by orcas – the drone filming clearly showed that the fin whales roll on their side when feeding, so their huge pectoral fins were breaking through the water in place of the much smaller dorsal fin.

Rob had once again decided to film the sequence while free-diving. However, even without the cumbersome scuba gear, bait balls move incredibly quickly, making it exhausting for the team, as Rob describes:

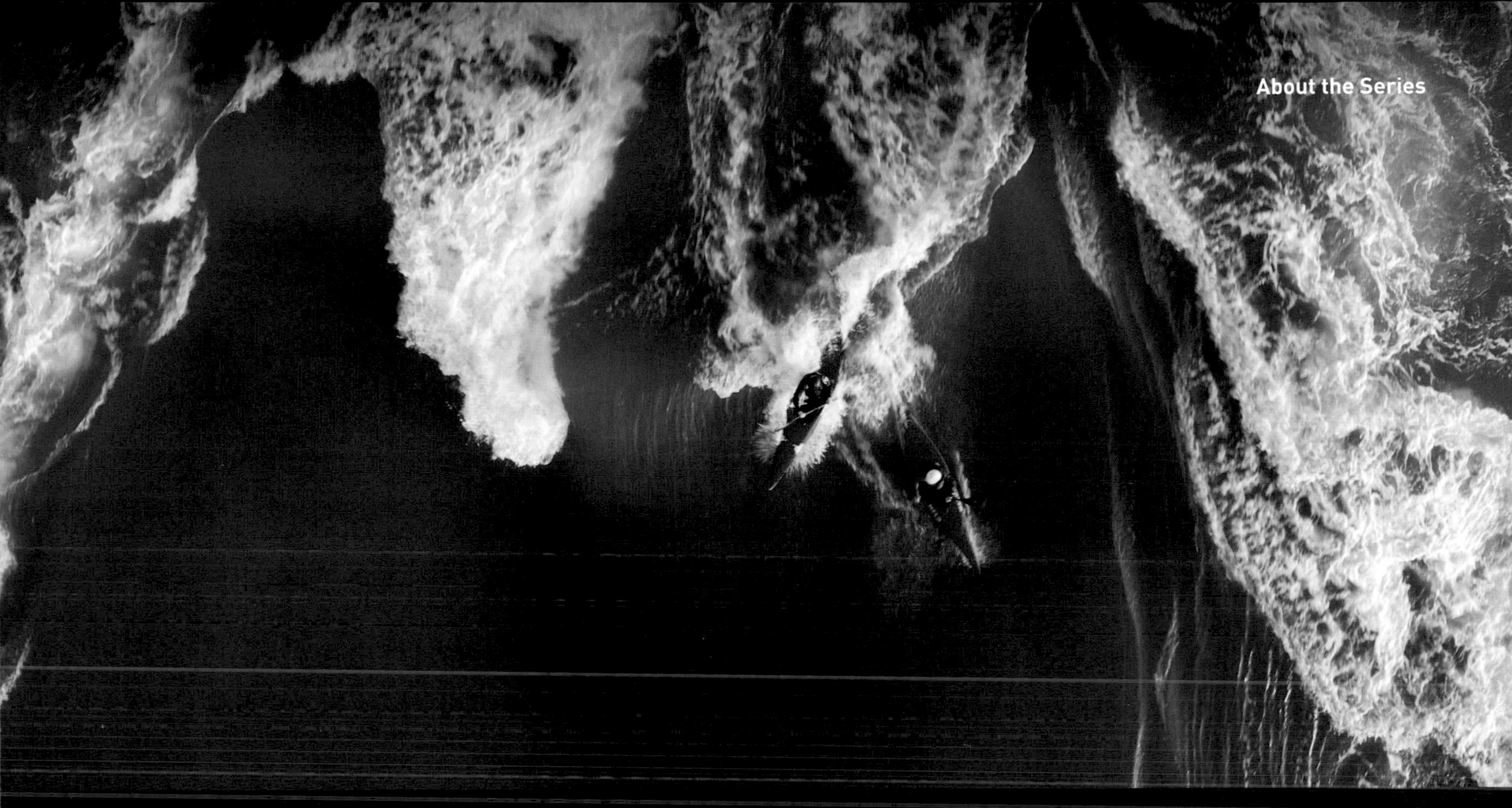

'You just tried to keep your heartbeat down. The adrenaline was really pumping, but you had to keep swimming as hard as you could to keep up with the bait ball. Next, the dolphins started swimming close by. 'Below them were massive tuna, also on the hunt for the fish. And then suddenly the whales started coming through very close.' With their eyes on dinner these huge animals didn't care about the strange small humans joining the fray: 'At one point Nicki touched me and pointed down and there was a fin whale pretty much swimming between my legs. It seemed to go on forever. At one point I could have touched it. We just chased around the bait balls all morning; some you won and some you lost.'

Rob and all the crew remarked on how small it made them feel filming the bait balls in the Celtic Deep:

'You are absolutely not the top of the food chain, you're just part of the food chain. Although there's nothing in there that would eat you, you are just a part of this massive predatory frenzy. It was insane that was all happening in British waters. People go to film the sardine run in South Africa, yet it's ridiculous how much is going on right here. It went on all day, it's absolutely crazy. I bet very few people have swum into a bait ball in British waters like that, and oh my God, we got the best sequence ever!'

Above: The use of kayaks, canoes and inflatables allowed the team access to the wildlife that would have been impossible by boat.

Coasts:
Life on the Edge

Bracing wind and waves from the Atlantic at Freshwater East, in the south-west of the Pembrokeshire National Park.

Jewel anenome.

Surrounded by sea on three sides, Wales is home to an amazingly diverse range of marine wildlife, from brightly coloured anemones and starfish to ocean-wandering seabirds to highly sophisticated marine mammals such as the harbour porpoise and bottlenose dolphin.

Whitesands Bay and Ramsey Island from the summit of Carn Llidi on the north Pembrokeshire coastline.

Facing the Atlantic
Exposed throughout the year to the often fierce Atlantic weather, Wales' coastline has been shaped by powerful and dynamic forces – winter storms can bring 10-metre-high waves and winds of up to 80 mph. Over the millenia the relentless assault of wind and waves has sculpted and reshaped the coast, creating spectacular rock formations. While ancient hard rocks project into the sea as rugged headlands, younger, softer rocks have yielded to the elements and eroded away.

Along much of the coast geological forces originating from deep within the earth have fractured and buckled the rocks, while fluctuations in sea level have dramatically shifted the boundary between sea and land.

But the shoreline provides vital food and protection too. In between the steep cliffs, vast dune systems and long sandy beaches form a much gentler coastal landscape; where rivers meet the sea – forming sheltered estuaries – the constant ebb and flow of the tide has created huge expanses of nutrient-rich mudflats and saltmarsh.

Biodiversity
The wide variety of habitats provides a safe haven for a wealth of plants and wildlife, including large colonies of Atlantic grey seals and the biggest semi-resident population of bottlenose dolphins in Europe.

In summer thousands of guillemots, puffins, Manx shearwaters and gannets gather to nest on rocky cliff fortresses high above the sea.

In winter secluded estuaries provide vital feeding grounds for tens of thousands of migrating wetland birds, which fly in to join the resident birds feeding on the tiny shellfish and worms hidden beneath the intertidal mudflats.

Celtic Myth and Legend

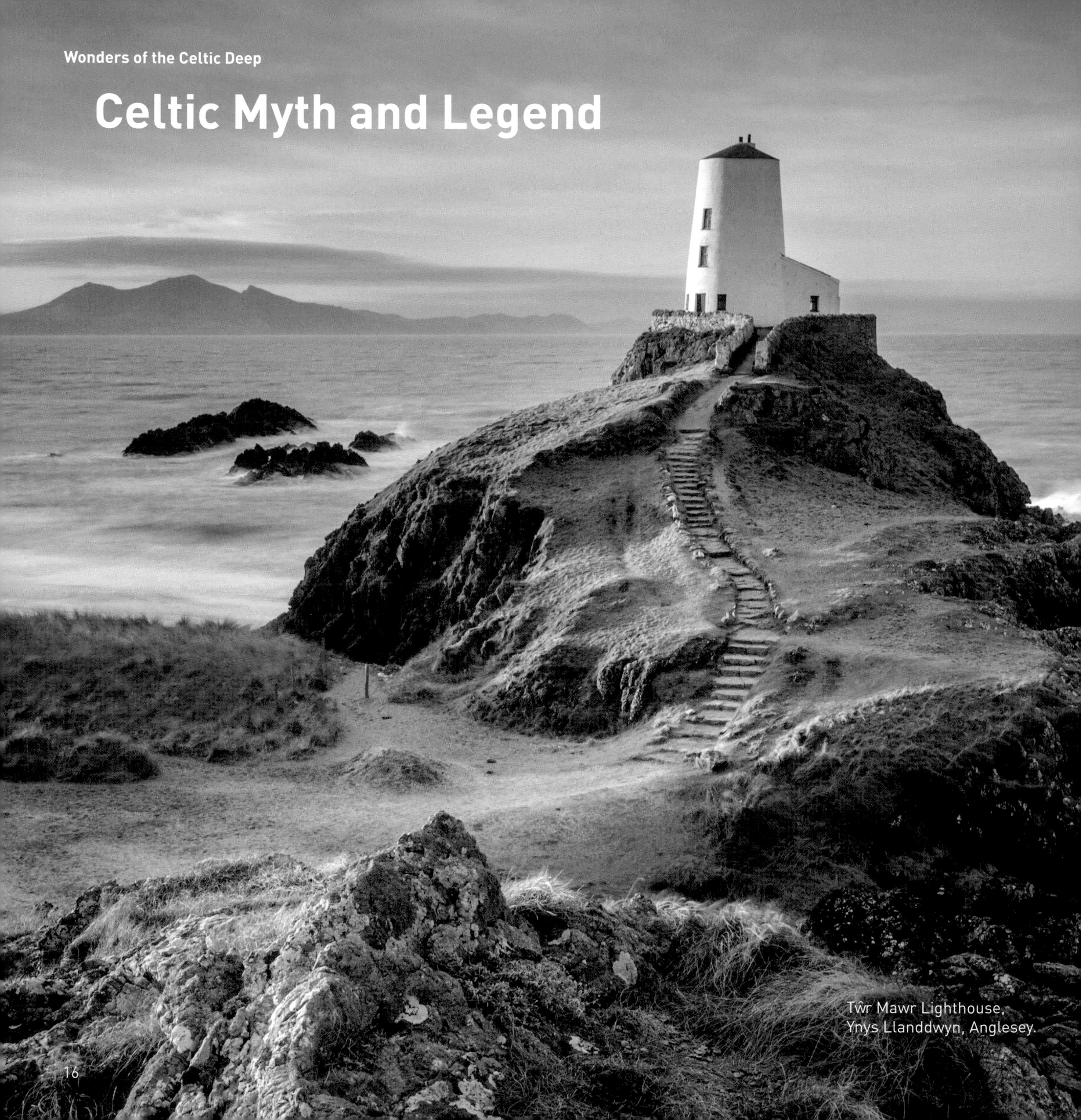

Tŵr Mawr Lighthouse,
Ynys Llanddwyn, Anglesey.

Facing towards the vast expanse of the Atlantic Ocean, the Welsh coastline is imbued with Celtic myth and mystery: a place where choughs are said to embody the soul of King Arthur, ravens are harbingers of death, seals can shift between human and animal forms and the boundary between the living and the mysterious otherworld that lies beyond can vanish, allowing ancient spirits to pass through.

The link between spirituality and this remote, windblown shoreline continued into early Christian times when pilgrims travelled to isolated cliffs and tidal islands to contemplate and connect to nature. Many of these spiritual retreats are still evident; scattered along the coastline are numerous holy wells and tiny churches and chapels such as Llanddywn on the far west tip of Ynys Môn (Anglesey) – dedicated to St. Dwynwen, Welsh patron saint of lovers – and St. Govan's chapel – a hermit's cell built into the side of a steep limestone cliff.

At certain times of year spectacular natural phenomena add to the palpable sense of magic and mystery. Following a summer heatwave flashes of electric-blue light illuminate the night sea. The shimmering display is a form of bioluminescence created by huge blooms of tiny plankton and algae emitting millions of photons of light.

Top right: In Celtic myth, people, places and nature are closely entwined. According to Welsh legend the red-legged chough incarnates the spirit of King Arthur.

Right: The 16,000-mile-long coastline provides food and shelter for an abundance of wildlife.

Moon and Tides

The mouth of the Burry Inlet. On a spring tide, 1,400 million cubic metres of water enters and leaves the estuary.

One of the most powerful and mysterious forces influencing the coastline is the moon. As the earth rotates on its axis, differences in the gravitational pull of the moon and the sun interact with the centrifugal force created by the spinning earth, causing the tide to rise and fall. Twice a day land becomes sea and sea becomes land.

The south coast of Wales has one of the biggest tidal ranges in the world – in some places up to 14 m – and results from the funnel shape of the Bristol Channel, which squeezes the incoming tide into an increasingly narrow waterway.

The twice-daily ebb and flow of seawater creates some of the most dynamic and extreme habitats on earth – rockpools. Despite the constantly changing environment, these are home to a rich array of wildlife. Capable of living half in and half out of the sea, rockpool animals show some of the most remarkable adaptations to the harsh conditions accompanying life on the edge.

The difference between high and low tide varies across the month and depends on the relative position of the earth, moon and sun. The largest tides – known as spring tides – occur when all three are horizontally aligned. The biggest spring tides of all take place around the time of the spring and autumn equinox. As the tide ebbs and flows it creates immense watery landscapes, neither land nor sea. On a fine March morning, as the sea re-enters the Burry Inlet, the tidal waterway lying between Gower and the Carmarthenshire coastline, the magical sea light transforms into a mirror, reflecting the vast, open sky.

Atlantic Grey Seals, Pembrokeshire Coast

The seas surrounding the Pembrokeshire coastline are home to one of the rarest seals in the world – the Atlantic grey seal. Around 5,000 live here, attracted by the rich fishing grounds and plentiful remote beaches and islands where they can rest, mate and give birth to their pups. On a summer's day they can often be spotted contentedly sunning themselves on rocky outcrops just offshore.

The largest marine predator in Wales, grey seals are skilful hunters. As well as having excellent hearing, their large round eyes are specially adapted to the low light levels found underwater; as they get closer to their prey, their highly sensitive whiskers then home in on their target. Their main diet consists of bottom-dwelling fish, such as sand eels and flatfish, but they also forage for crabs, lobster and octopus.

The Smalls, an outcrop of sea-washed rocks 20 miles off the Pembrokeshire coast, is a favourite haul-out spot for Atlantic grey seals.

With their sleek, streamlined bodies and large flippers, seals are powerful swimmers. While some rarely stray far from the Pembrokeshire coast, others journey hundreds of miles to forage for food; some individuals have been discovered as far afield as northern Spain and the west coast of Ireland.

Seals are also master divers. Descending on a single breath of air to depths of up to 70 m, they can stay underwater for up to 30 minutes at a time. Unlike humans, they expel air from their lungs before diving, relying on oxygen stored in their blood and muscles to sustain them throughout each dive. A thick layer of blubber and waterproof double fur coat helps keep them warm in the cool Welsh waters.

In between hunting forays the seals come ashore to rest or 'haul out'. Although they live mainly solitary lives, on a sunny day large groups of seals may congregate on the best-situated rocky outcrops and beaches. Male seals are known as bulls and are distinguishable by their broad, convex, 'Roman' nose, thick neck and finely mottled dark fur, while the females, known as cows, have slender heads and are typically silver–grey with a pale underside, patterned with dark spots and blotches. Each seal has its own unique markings which remain constant throughout its life.

Top: Seals regularly take short naps in the sea, bobbing about vertically in the water, a behaviour known as 'bottling'.

Right: The *Celtic Deep* team captured this charming pair of young seals playing together in the shallow coastal waters.

Annual Moult

Above: A group of seals gather to moult on an isolated beach beneath Cemaes Head, the highest sea cliff in Wales. According to local legend, a mermaid once lived in the surrounding seas – it's possible the story is linked to the eerie calls of seals often heard echoing around the Pembrokeshire coastline.

Between January and April, the seals haul out on remote, isolated beaches for the annual moult. After enduring months of harsh winter weather, this is a difficult time for the seals as they won't eat until they have replaced their old fur, a process which can take up to six weeks. Being on top of each other for such a prolonged length of time leads to some grumpy exchanges, particularly as moulting is usually an itchy process. As soon as they have grown their new coat the seals return to the sea to feed on the shoals of fish migrating into the warming coastal waters. By now the females are around three months pregnant and must gain as much weight as possible in preparation for giving birth later in the year.

FACT FILE
ATLANTIC GREY SEAL

- Britain's largest marine predator.
- - Males: around 2 m long. Weight: 230–300 kg.
  - Females: around 1.8 m long; weight: Around 155 kilos.
- **Flippers:** 25 cm long.
- **Daily food intake:** about 5 kg a day.
- - Males: Convex Roman nose, dark in colour.
  - Females: Slender head, silver-grey fur and pale underside with dark markings.

Seabird City on South Stack, Anglesey

Above: South Stack, Anglesey in early June. Around 11,000 guillemots nest on the steep cliffs.

Guillemots pair for life. Males and females take it in turns to incubate the egg on their feet.

For a few months each year, the steep, barren cliffs at South Stack, on the island of Anglesey, are transformed into a crowded seabird city as thousands of guillemots and razorbills arrive to nest and rear their young.

Guillemots spend most of the year out at sea but each spring they return to the same tiny ledge. Adults pair for life (with the odd fling), affirming their bond with lengthy bouts of preening. The female lays a single pear-shaped egg, which the couple then take turns to incubate. Guillemots are renowned for their clumsiness and the egg is shaped and weighted to reduce the risk of the parents inadvertantly knocking it off the ledge.

Guillemots are highly sociable and neighbouring pairs often form close friendships, but living on a cramped ledge can be stressful and quarrels frequently break out. Neighbouring birds also spend time grooming each other, which helps reduce tension in the colony.

Communal living brings distinct benefits. Throughout the day predatory gulls and ravens patrol the cliffs, on the lookout for eggs to feast upon. With their backs turned to the sea, the guillemots form a protective wall, shielding their eggs from attack. Great black-backed gulls are particularly ruthless predators. Swooping down from on high, they try to panic the guillemots into abandoning their eggs. These are the largest gulls in the world and not many guillemots can withstand a direct assault. Each couple must be constantly on their guard if their precious egg is to hatch successfully.

Above: Predatory ravens and gulls patrol the cliffs, on the lookout for eggs and chicks. Living in a densely packed colony provides safety in numbers, with neighbouring guillemots joining forces to fend off attacks.

FACT FILE
GUILLEMOTS

- Guillemots belong to the auk family, which includes puffins and razorbills.
- Around 11,000 guillemots nest on the steep cliffs at South Stack.
- Each pair occupies a patch of rock only a few centimetres wide, the smallest nest size of any Atlantic seabird.
- Both sexes have chocolate-brown plumage on top and white below.
- The distinctive bright blue eggs are laid from mid to late May, hatching around 30 days later.
- Like most seabirds, guillemots are long lived – the average lifespan is 27 years.

Silver-Studded Blue Butterfly, South Stack

South Stack is one of the few remaining sanctuaries in Wales for the beautiful silver-studded blue butterfly – now increasingly rare due to the disappearance of maritime heathland from along much of the coast.

The butterflies get their name from the distinctive row of metallic studs on the underside of their wings. While the males' wings are a soft powder blue, the females are light brown in colour. They live in small, tightly knit colonies, feeding on the low-lying, nectar-rich flowers, with the females rarely straying more than 20 m from their home patch. Their most important food plants are bell heather, gorse, heather and cross-leaved heath, which thrive on the windblown cliffs.

From late June to August, male butterflies can be seen hovering low above the brightly coloured patches of gorse and common heather looking for females. Once a male finds a receptive female, he joins the tip of his abdomen to hers and then deposits a tiny packet of sperm and nutrients which the female uses to fertilise her eggs. After mating the female lays the eggs one at a time low down on the shrubby stems of heather, which provide much-needed shelter from the ferocious winter storms.

The larvae hatch out in the spring, feeding on the growing tips and flowers. As well as providing shelter and food, the heathland also supplies another key ingredient for their survival – colonies of black ants. These care for the young caterpillars, guarding them until they emerge as adult butterflies. It's thought they pick up the larvae soon after hatching and place them in tiny chambers beneath rocks or stones. In return, the caterpillars supply the ants with sugary foods. This remarkable relationship is so important that the female silver-studded blues will only lay their eggs where they can detect the presence of black ants.

Changes in land use have led to the loss of 80 per cent of UK maritime heaths and with them the demise of this lovely butterfly. Fortunately for the silver-studded blues, the RSPB is now actively managing this precious lowland heathland, creating the mosaic of long and short vegetation they need to sustain them throughout their complex and intriguing life cycle.

Left: From late June to July the sea cliffs are carpeted in colourful heathers and gorse.

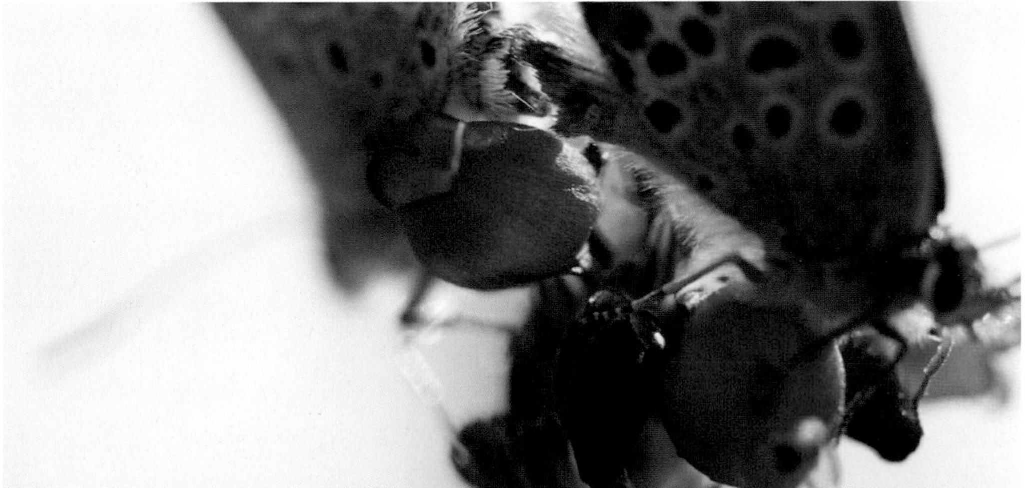

Above: After hatching, the young caterpillars are dependent on black ants, which tend them until adulthood.

Joining the tips of their abdomens, the male deposits a package of sperm and nutrients which the female uses to fertilise and nourish her eggs.

FACT FILE
SILVER-STUDDED BLUE BUTTERFLY

- Silver-studded blue butterflies can be seen at South Stack between June and August.
- Key food plants are bell heather, gorse, heather and cross-leaved heath.
- Males are powder blue in colour while the females are light brown. Both sexes have a row of metallic studs on their wings.
- The female usually lays a single brood of eggs over a period of days. These then overwinter until the caterpillars hatch out in spring.
- Over the past 150 years, changes in land management have led to the disappearance of 80 per cent of maritime heathland.

Shelduck Nursery on the Teifi Estuary

Dippers.

As the River Teifi approaches Cardigan Bay it transforms into a narrow tidal estuary. Surrounded by a mosaic of woodland, reedbeds and tidal mudbanks, the sheltered waterway supports a wealth of wildlife.

As it winds through Cilgerran Gorge and approaches Cardigan Bay, the River Teifi becomes increasingly influenced by the tides. Twice a day, as the sea sweeps in, it mixes with the fresh water flowing downriver.

Once an important trade route for the historic coastal settlements of Cardigan and St Dogmaels, this secluded waterway is a haven for wildlife throughout the year. In autumn, salmon and sewin migrate in from the sea to spawn upriver, while in winter the surrounding marshes provide food and shelter for flocks of wigeon, lapwings and curlews.

In spring and summer the wildlife changes again as the banks of the estuary become a nursery for a wide range of birds, including a small population of colourful shelducks. These large, handsome ducks are easily recognisable by their distinctive bottle-green heads, broad chestnut breast bands and bright red bills.

In spring, pairs of males and females move into the surrounding fields to nest. Shelducks are one of the few water birds to regularly lay their eggs in burrows. Rabbit holes are a firm favourite, but they may also nest in tree cavities, outbuildings and even upturned boats. Thirty days after the chicks hatch their parents lead them down to the estuary. Here they spend the next few weeks feeding on the plentiful water snails and other invertebrates living beneath the surface of the mudbanks.

Left: In early summer the Teifi estuary is home to families of shelducks, which feed on the nutrient-rich mudbanks.

Shelducks have large clutches – around eight to ten eggs – this is a key survival strategy as not all the hatchlings will make it to the safety of the water. These tiny balls of fluff are easy pickings for predators such as foxes and gulls, which actively track them on their perilous journey down from the surrounding countryside. Around two-thirds of the ducklings are lost in the first two weeks of life.

On reaching the estuary the risk of predation remains high and the parents must stay on guard at all times. If either parent spots danger they attempt to draw attention away from the youngsters by swimming across to the opposite bank, calling out to their offspring to take cover.

As further protection against predators some pairs join forces with other parents and non-breeding birds, gathering their chicks together into large crèches. By mid-June, groups of ducklings of various ages may be seen bobbing along the estuary.

Towards the end of June the adult birds leave their offspring, flying far away to larger, safer waterways where they moult their feathers and develop their winter plumage. By this time the young shelducklings have grown considerably and are much better equipped to fend off predators. At seven weeks old they are able to fly; they too then disperse over winter before returning in springtime, when they often help out with babysitting the new generation of shelducklings.

Newly hatched shelducklings are highly vulnerable to predation by foxes. Up to two-thirds of chicks die in their first two weeks of life.

Towards the end of June the adult shelducks fly off to moult, leaving their adolescent offspring to fend for themselves.

FACT FILE
SHELDUCKS

- Adult males and females are predominantly white, with bottle-green heads, a broad chestnut breastband and a bright red bill.
- Males are larger than females with a much more prominent beak shield.
- Length: 58–65 cm.
- Juvenile ducks are white with grey-brown heads, backs and wings.

Bottlenose Dolphins, Cardigan Bay

Cardigan Bay is home to one of the wildlife stars of Welsh seas – the bottlenose dolphin. Dolphins can be seen in the bay throughout the year, but the best time to spot them is in summer when, attracted by the increasingly plentiful shoals of mackerel, females move closer to the coastline to feed and care for their young.

The dolphins travel around the bay in small groups or 'pods'. They hunt for their food by echolocation: by producing clicking sounds and interpreting the returning echoes they can tell the size, shape and even the speed of their prey. Mothers pass their hunting skills down to their calves; there's a lot to learn and the youngsters won't become fully independent for between three and six years. Male dolphins tend to move about in smaller separate groups, but family ties remain strong, and they too may sometimes be seen swimming alongside the mothers and calves.

Dolphins are renowned for their sophisticated social behaviour, using a variety of clicks, squeaks and whistles to communicate with each other. Each dolphin has its own unique signature whistle – it's thought this serves the same purpose as a human name, enabling individuals to keep in touch with each other even when separated. Intriguingly, recent research indicates the Cardigan Bay dolphins may have their very own dialect – their whistles are a higher frequency than those recorded anywhere else in the world.

In between foraging for food the dolphins engage in another favourite activity – playtime. This involves a wide range of behaviours, including leaping out of the water or 'breaching', tail slaps and bow riding.

At times these playful activities can turn into much rougher knockabouts – a closer look at the dolphins' fins reveals numerous scars and notches. The *Celtic Deep* team captured another example of the darker side of dolphin behaviour when, on the last day of filming, a group of young males swam up to a family group and started harassing the females in an attempt to mate with them.

New Quay, Cardigan Bay. In summer, bottlenose dolphins are frequently seen hunting for fish in and around the harbour.

Right: The dolphins criss-cross Cardigan Bay in small family groups, constantly communicating with each other via their signature clicks and whistles.

FACT FILE
BOTTLENOSE DOLPHINS

- In summer, Cardigan Bay is home to around 300 bottlenose dolphins – the largest semi-resident population in Europe.
- Dolphins feed on a wide variety of fish, squid and shellfish, tracking down their prey using echolocation.
- Bowriding enables the dolphins to travel twice as fast for the same energy cost.
- Individuals can be identified and monitored by the pattern of scars on their fins.

South Stack Guillemots

It's mid-June at South Stack. By now most of the guillemot chicks have hatched out and the noise on the cliffs is truly cacophonous as their parents fly to and from the sea in search of fish for their demanding offspring.

Feeding the chicks is a full time job so their parents take it in turns to forage out at sea. Before embarking on a fishing trip they often take a quick bath at the bottom of the cliffs – they've been on South Stack since April and the cramped ledges are now soiled and smelly. Glands beneath their wings secrete a powerful waterproofing oil to keep their feathers in prime condition. The bottom of the cliffs is also a good vantage point to observe from which direction the successful guillemots are returning, enabling them to judge where the best diving spots are likely to be.

Fishing trips can be lengthy – up to 30 km and back. To the human eye the sea surface can look much the same. But the guillemots are able to pick out subtle variations in colour and so identify fast tidal currents and upwellings of where the most fish are to be found.

Guillemots are superb divers – on a single breath of air they can descend to an astonishing 70 m. Flapping their short wings they literally fly through the water, pursuing their prey in a high speed chase. Their main foods are sand eels and sprats, which they catch one at a time, carrying them back head first to the waiting chick.

Returning to the cliff, another challenge faces the parents – how to find their chick on the crowded ledges. Amazingly, despite the tremendous din, both chicks and parents can distinguish each others' calls. The chicks need to mature quickly – in just three weeks' time they will leave their high-rise nursery to start a new life out at sea.

Above: By mid-June most of the guillemot chicks have hatched out, turning the cliffs into a gigantic avian nursery.

Parents take it in turns to embark on lengthy fishing trips.

It's important to keep their waterproof plumage in tip-top condition. Before diving they often take a quick dip to clean their feathers.

FACT FILE

GUILLEMOTS

- Guillemots can travel up to 30 km in search of fish.
- Their main prey are sandeels and sprats.
- Diving up to 70 m, they can stay underwater for over 4 minutes.
- Guillemots search for upwellings and fast tidal currents, as this is where most fish are likely to be found.
- Conservationists fear South Stack's guillemot colony could be badly impacted by the construction of tidal energy turbines, as they are attracted by turbulent water.

Rockpool Dramas in Fishguard Bay

As the tide recedes at Fishguard Bay it reveals a patchwork of rockpools containing a huge variety of wildlife.

As the tide rises and falls, seawater collects in hollows within the rocky outcrops scattered across the beach, creating a mosaic of rockpools. Home to an abundance of wildlife, these are one of the most magical – and extreme – of all coastal habitats.

Life in a rockpool can be tough. In the course of a day the animals are exposed to massive changes in their environment, from crashing waves to huge swings in temperature, salinity and humidity. The extent of these fluctuations varies, with rockpools on the upper shore experiencing the greatest extremes.

But the incoming sea also brings a regular supply of food and oxygen and – for those animals that can cope with the changeable conditions – shelter. As the sea retreats, a period of calm descends and one by one the inhabitants emerge from their hiding places, transforming the rockpool into a bustling metropolis.

As in any busy community, competition for food and space can be fierce – in such an enclosed space it's all too easy to become someone else's dinner. Starfish are one of the rockpool's top predators, using their acute sense of smell to detect their prey. They may appear immobile, but on the undersurface of each of their arms are hundreds of extendable tube feet which they employ to creep up on their prey. Mussels are one of their favourite foods. Gripping the mussel's shell with tiny suckers on the tips of its tube feet, the starfish prises it apart; turning its stomach inside out it then pours digestive enzymes over the soft, fleshy creature sheltering inside before scooping up the contents.

Right: Competition for food is fierce. Here a common blenny and a group of prawns battle it out for possession of a sea snail.

Common Blenny: A Fish Out of Water

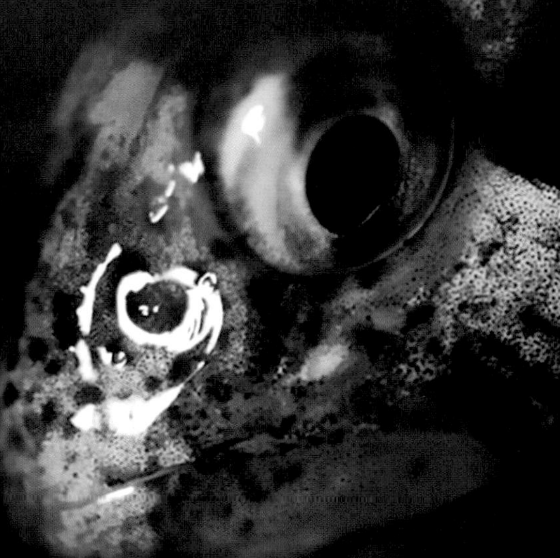

Unlike most fish, the common blenny can breathe through its skin, enabling it to survive for long periods out of water.

FACT FILE
COMMON BLENNY

- Blennies are feisty little fish. When disturbed, their sharp teeth can give a nasty nip.
- Instead of scales, a thin coating of slime enables blennies to breathe through their skin.
- Living for up to 20 years, they rarely stray from their chosen rockpool.
- Usually a mottled grey-brown, but in the breeding season the male becomes black with bright white lips.
- Breeding takes place from April to August. Once the female has laid her eggs, the male takes over parental duties, guarding them until the young hatch out.

The common blenny, or 'shanny', is one of the most extraordinary of all rockpool inhabitants. At mid-tide they dart around the pool making the most of the plentiful food on offer: ragworms, snails and lugworms are the mainstay of their diet, along with barnacles, which they bite off the rocks with their strong teeth.

Blennies are highly territorial – once they've found a good rockpool they rarely stray far from home.

Most intertidal animals struggle to survive for prolonged periods out of water, but the common blenny has two remarkable features which enable it to hold out in its home patch during even the biggest of spring tides. A protective layer of mucous means that, unlike most fish, they can breathe through their skin as well as their gills, but for this slimy coat to function they must stay moist. Their second extraordinary adaptation enables them to do just that. As the sea ebbs away they prop themselves up on their fins and slowly propel themselves into a deep, damp crevice, where they wait it out until the sea returns.

At low tide this rockpool becomes a potential deathtrap – the stranded blenny must find a damp crevice as soon as possible.

Burry Inlet Cockles

Llanrhidian Marsh, on the shores of the Burry Inlet. As the tide ebbs and flows, freshwater from the River Loughor mixes with the incoming sea.

Lying between north Gower and Carmarthenshire, just a few miles from the busy M4 corridor, the Burry Inlet and Loughor estuary form one of the most dramatic and dynamic coastal landscapes in Wales – on a spring tide, the difference between high and low water can be nearly nine metres. As the sea retreats it exposes a vast watery wilderness, made up of glistening sandbanks, mudflats, saltmarsh and twisting tidal creeks. With its fast-rising tides and treacherous currents, the estuary was once notorious for shipwrecks, but for hundreds of years the constantly shifting shoreline has provided an important source of livelihood for local people – cockles.

Flooded by the sea twice a day, the sand and mud are crammed with these small, edible molluscs. Easily recognised by their heart-shaped shells, they lie just a few centimetres beneath the sand, filter-feeding on the abundant microscopic plankton carried in on the tide. Generations have made a living from this seaside larder – in the early 1900s, around 250 gatherers came here each day to extract the cockles, transporting them back to the shore by donkey and cart. Today a small but thriving cockle industry still exists. While the donkeys

have long been replaced by tractors, the back-breaking traditional method of rake and riddle still continues – a skilled cockler can harvest thousands in a day.

Top: Whiteford Lighthouse was built in 1865 to help ships navigate the treacherous shoreline.

Right: Today cockles are still gathered by hand, rake and riddle, as they have been for generations.

FACT FILE
BURRY INLET

- Covering over 9,500 hectares, the Burry Inlet and Loughor estuary is the largest estuarine complex entirely in Wales.
- At low tide the exposed sand and mud stretches for seven miles.
- Cockles feed on microscopic plankton floating in on the tide.

Burry Inlet Oystercatchers

A winter dawn on Whiteford Sands on the Gower Peninsula. As the tide recedes, a flock of oystercatchers leave their night roost in the dunes and congregate along the shoreline to feed.

Under the glittering mud and sand lies a sumptuous banquet made up of tens of thousands of tiny crustaceans and worms. The oystercatchers may look as if they are randomly probing the sand but extracting this rich seafood platter is skilled work. In addition to hunting for visual clues, small sensors at the tips of the bills act as natural metal detectors enabling them to locate their prey beneath the surface of the sand.

From December to March the Burry Inlet's resident population of around 6,000 oystercatchers rises to over 14,000 as winter visitors fly in from Iceland and northern Scandinavia in search of food and shelter. These vital feeding grounds are shared with many other wading birds, including curlews, knots and dunlin. The oystercatchers can be easily recognised by their striking black and white plumage, bright orange beak and legs and noisy calls. Throughout the day their constant high-pitched 'kleeps' pierce the cold winter air.

As the tide turns the oystercatchers retreat to their daytime roosts along the Burry Inlet and across to Swansea Bay. They will return to feed at the next low tide before retreating to the dunes to sleep.

During the night the oystercatchers are vulnerable to predation by foxes. To make it easier to escape they usually roost in a long line – if one oystercatcher senses danger, it emits a loud alert call to warn its neighbours and they all take flight together.

Above: In winter the Burry Inlet is home to around 14,000 oystercatchers. As the tide recedes they gather along the shoreline to feed on the tiny shellfish lying just below the surface.

Oystercatchers are very vocal, constantly calling out to each other with loud 'kleep kleeps.' Picky eaters, they often wash their prey before downing it whole.

FACT FILE
OYSTERCATCHERS

- Large wading bird, 40–45 cm long.
- Black and white plumage with cherry red beak, legs and eyes.
- Gather in sheltered estuaries and inlets, sharing their roosting and feeding grounds with other waders, such as curlews, knot and dunlin.
- Oystercatchers' favourite foods are cockles, mussels, worms and whelks.
- It can take an oystercatcher two to four minutes to eat one cockle.
- Average lifespan – 12 years.

Ynyslas: An Ancient Drowned Forest

At low tide, strange shapes start to emerge across Ynyslas beach on the west coast of Wales. A fierce storm has scoured out the sand to reveal hundreds of fossilised tree stumps; oak, pine, birch, willow and hazel. This petrified forest dates back around 4,500 years to the end of the last Ice Age, when sea levels around Britain were much lower.

As temperatures rose and the ice sheets retreated, the encroaching sea drowned large areas of coastal Wales. Soon the forest became preserved under a thick blanket of peat.

After stormy weather, at low tide hundreds of petrified tree stumps, the remains of an ancient forest, resurface at Ynyslas beach.

The drowned forest dates back to the last Ice Age, when sea levels were much lower.

Archaeologists have discovered tantalising clues that these drowned lands were once important hunting grounds for Mesolithic and Bronze Age hunters – fossilised human and animal footprints as well as simple tools made from antlers have been discovered in the peaty deposits surrounding the tree stumps.

The ancient forest is thought to have inspired the legend of the mythical sunken kingdom Cantre'r Gwaelod, the lost Atlantis of Wales. According to the legend, the kingdom extended some 20 miles west off the current shoreline into what is now Cardigan Bay, but was lost to the sea when Seithenyn, the guardian of the sea defences, had too much to drink and forgot to close the floodgates. It's said that if you listen closely, you can hear church bells ringing out from under the waves.

These ancient fossilised forests are a reminder of how the coastline has changed over time and how, as climate change causes sea levels to rise, long stretches of coastal Wales could one day be swallowed by the sea.

FACT FILE

THE PETRIFIED FOREST

- Located five miles north of Aberystwyth along Borth and Ynyslas beaches and best seen at very low tides or after a winter storm.
- The drowned forest probably flourished for at least 1,000 years before the sea level rose.
- The tree stumps were preserved by a thick layer of peat.

Kittiwakes of Mumbles Pier

Each summer around 100 kittiwakes nest under the old lifeboat station on Mumbles Pier. Once the chicks have hatched out, their parents share feeding duties, hunting on the wing for small fish.

FACT FILE
KITTIWAKES

- Long-distance flyers, kittiwakes spend most of the year far out in the Atlantic Ocean.
- Length: 38–40 cm.
- Diet: Small fish, shrimps and worms.
- Dark eyes, yellow bills, silver-grey backs, black-tipped wings, and loud 'kittiwaaake' calls.
- Nest building begins in April, with two to three eggs being laid in May.
- Incubation period: 27 days.
- The chicks are fed by both parents for 40–45 days, only leaving the nest when they are strong enough to fly.

All seabirds must negotiate the two contrasting worlds of land and sea, but kittiwakes have taken ocean living to new limits. These elegant little gulls spend most of their lives hundreds of miles out in the Atlantic Ocean, only returning to dry land in spring to nest and rear their young. With their silver-grey backs, soft, dark eyes and long, black-tipped wings, they are one of the most captivating of all seabirds.

Kittiwakes usually construct their nests on narrow ledges on steep, high cliffs, but one of the best places to see them is at the end of Mumbles Pier, nesting on specially built platforms underneath the old lifeboat station. On a summer's day the kittiwakes can be easily identified by their loud signature 'kittiwaaake' calls as they circle high above Swansea Bay.

Adult kittiwakes start nest building in April, creating a shallow bowl made from seaweed, grass and mud in which the female lays two to three eggs. The chicks hatch out around 27 days later and are looked after by both parents, who take it in turns to hunt for sandeels and other small fish. Unlike the short-winged, deep-diving guillemots, kittiwakes' long wings are adapted for long-distance flying – hunting on the wing, they swoop down to pluck their prey from just below the sea surface.

The growing chicks develop a similar colour plumage to their parents but can be distinguished by their black neck collar, black 'W'shape across their wings and thick black tail band. The youngsters will not leave the nest until they can fly. In August they and their parents migrate far out into the Atlantic, where they must quickly learn how to survive the harsh conditions out at sea.

Grey Seal Pupping, Pembrokeshire

In autumn, Pembrokeshire's rocky coastline is transformed into a sprawling maternity ward as female Atlantic Grey seals return to remote beaches and caves to give birth. At this time of year the haunting sound of seals calling out to each other can often be heard resounding along the shoreline. Pembrokeshire is one of the most important breeding areas in Wales – by the end of the year over 1,000 pups will have been born.

Each female gives birth to a single pup. Seals are attentive mothers and the survival rate is high – around 80 per cent. Storms are one of the main dangers facing the young pups and experienced females know to choose sheltered spots well above the high tideline to give birth. This is a vulnerable time for the mothers too – the actual birth is rarely seen and usually over within 10 minutes.

The newborn pups' coats are yellowy-white – a legacy of the Ice Age when this would have helped them blend into the snow and ice. Weighing just a few kilos and barely mobile, the pups are initially totally dependent on their mothers. But they must grow fast – within three weeks, their mothers will leave them behind and head back to sea. Feeding four to five times a day on their mothers' rich milk – an astonishing 60 per cent fat – the pups put on nearly two kilos a day. Within three weeks they will have trebled their birth weight.

During the first few days of life the mothers rarely leave their pups, but as they grow bigger and stronger the females start to take quick dips just offshore, where they can be seen bobbing about in the water, still keeping a watchful eye on their offspring.

Within a week most pups are tentatively beginning to explore the sea. Looking down from the Pembrokeshire coastal path you may be lucky enough to see groups of pups playing in the surf. This behaviour is not without risk – the pups' swimming skills are still very basic and they cannot withstand rough seas; if they become separated from their mothers they will not survive for long. As the pups become more active the beaches become increasingly noisy as they call out to check their mother's whereabouts.

Within three weeks the pups' appearance has changed radically as their fluffy white fur is replaced by a darker, more waterproof coat in preparation for life at sea.

Left: Newborn pups have yellowy-white coats – a legacy from the Ice Age when this would have provided excellent camouflage on the snow and ice.

Seal milk is an astonishing 60 per cent fat – by the time the pups are weaned three weeks later they will have trebled their weight to around 45 kilos.

By now they have begun to develop the all-important layer of blubber which will help insulate them from the cold winter seas.

The pups benefit from their mothers' full protection for two to three weeks – even before they are fully weaned male seals have already started to crowd onto the beaches in search of mates. Fights between rival males can be brutal, making it a dangerous time for the young pups, who are at risk of getting crushed on the increasingly cramped beaches.

Once one male, the so-called 'beachmaster', has laid claim to all the females on the beach, he will guard them fiercely. Soon after mating the females abandon their pups and head out to sea to feed. It's important they build up their dwindling fat reserves before winter sets in – while their pups are growing fast, over the same period their mothers will have lost up to a third of their body weight. Implantation of the fertilised egg is delayed for three months, which means they have another summer to feed up before giving birth again, and guarantees that pupping occurs at the same time each year.

Separated from their mothers, the pups must quickly learn to fend for themselves; many stay on the same beach for several weeks, living off their fat reserves while they develop their hunting skills. Sadly, the increasingly fierce winter storms resulting from climate change inflict a heavy toll and not all will make it through to spring, but those that do have a good chance of enjoying a long life off Pembrokeshire's rugged coastline – grey seals can live up to 35 years.

Right: By the time they are just a week old many pups are already splashing about in the surf, but they are still very small and must rely on their mothers to get them out of trouble.

FACT FILE
GREY SEALS

- Female seals return to the same breeding area each year.
- Each female gives birth to a single pup, weighing around 14 kilos.
- After a couple of weeks the pups start to shed their baby fur, replacing it with a velvety waterproof coat.
- Females first mate at five to six years and can continue to breed for over 25 years.
- Competition between the bull males for females is fierce. Succesful males are likely to be 11–16 years of age.
- Three weeks after giving birth the females return to the sea, leaving their pups to fend for themselves.

South Stack, Jumpling Dramas

It's early July at South Stack. As the guillemot chicks grow in size and confidence, life on the crowded ledges is becoming even more chaotic and noisy. But all that is about to change. Over the next week the guillemot colony will undergo a dramatic transformation as, one by one, the chicks leave their clifftop nursery to start a new life out at sea. They are still very young – only three weeks old – their wings are still not fully formed and they cannot fly. At the bottom of the 70-metre-high cliff, jagged rocks project above the rough waves. This will be one of the biggest gambles of their entire lives.

As the chicks' fathers call out from below, the mothers gently coax their offspring closer to the edge. Unable to fly, the only way down is to jump. With South Stack's resident ravens and black-backed gulls still watching the chicks' every move, the parents wait until dusk before encouraging their precious offspring to take the plunge. Weather conditions can make all the difference between success and failure. Strong winds are particularly dangerous – there is only a narrow channel of relatively sheltered water at the bottom of the cliff – and once they launch themselves from the cliff edge the chicks have little control over where they land.

Even in good weather not all the chicks will survive – they have no prior experience to guide them and it's easy to misjudge the narrow landing pad.

The chicks' ever-attentive parents rush to the aid of those who fall short, but the force of the impact means that many will die.

Above: A guillemot chick prepares to launch itself off the cliff ledge – the drop to the sea is around 70 m.

Others reach the sea but then drown after being swept back onto the rocks by the rough waves and powerful currents swirling around the cliff.

Within three to four days most of the so-called 'jumplings' will have left South Stack. As more and more chicks launch themselves off the ledges, those left behind become increasingly agitated. While the *Celtic Deep* team were filming one stormy evening, the adult guillemots could be seen frantically trying to hold back their offspring who had rashly decided this was a good moment to jump.

For those chicks that do reach the safety of the water, this heart-stopping leap of faith marks the start of a totally different way of life. Although they are capable of diving as soon as they hit the water, they are not able to fly properly for a further two weeks. After reuniting with their parents, they swim out towards the open ocean. Over the next two to three months the father will pass on his all important fishing skills, before they both go their separate ways. The young guillemots will not head back to South Stack for at least another two years.

After the months of hectic activity, by the end of July South Stack falls strangely silent. The adult guillemots spend the winter foraging in the coastal waters, but return to the cliffs in small numbers, a few hundred birds at a time, from January, staying for a few hours. This happens every few weeks and the theory is that they may be non-breeding birds looking at potential nesting sites. Their number increases with each visit as the end of April approaches, when the whole colony will take residence again.

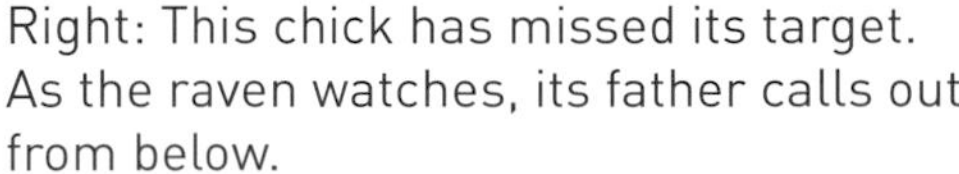

Right: This chick has missed its target. As the raven watches, its father calls out from below.

Coasts: Diaries
Bottlenose Dolphins

To capture the rich social behaviour of bottlenose dolphins, the *Celtic Deep* team travelled to Cardigan Bay, home to the largest semi-resident population of dolphins in Europe.

Bottlenose Dolphins

To create an immersive experience camera operators Rob Hill and Dan Burton wanted to film the dolphins from above and below water. But with only four days to complete the shoot, time was short. To help locate the dolphins the *Celtic Deep* team enlisted the help of Dr. Sarah Perry and local skipper Steve Hartley, who have been monitoring the Cardigan Bay dolphins for over 20 years. But even with their considerable expertise, success could not be guaranteed. Welsh weather is notoriously unpredictable and and the team needed calm seas and good visibility to locate and film the fast-moving dolphins.

With the drone camera capturing the dolphins from the air, the main challenge was to get close up shots of them swimming from just above and below water. Prior to the shoot, Rob and Dan had spent weeks designing a purpose-built polecam rig.

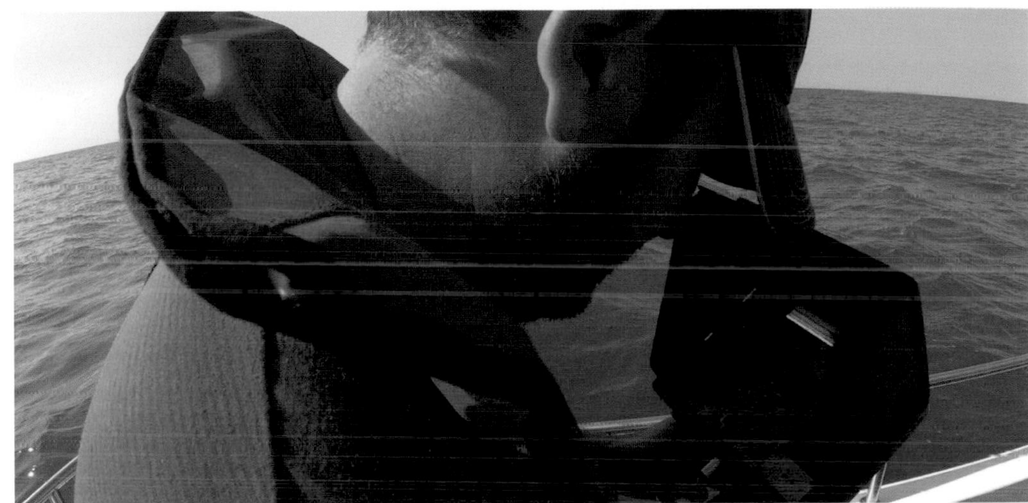

Left: The *Celtic Deep* team travelled to New Quay, Cardigan to film the dolphins.

Top: Dolphins can swim up to 18 mph. To keep the camera steady while filming at speed, the technical team designed a purpose-built pole rig.

Right: The team enlisted the help of marine biologist Dr. Sarah Perry and boat operator Steve Hartley of Cardigan Bay Marine Wildlife Centre to locate the dolphins.

Unfortunately, despite all the painstaking preparation, the shoot got off to an unpromising start with foggy conditions above water and murky seas below. The new equipment also proved trickier to operate in the field than anticipated – it took considerable trial and error to perfect the camera angles and at one point the lens completely fogged up. By the end of the third day of filming, spirits were getting low. While the team had been able to get some striking aerial footage of a family group foraging in the bay, they still had comparatively few underwater sequences and none of the rich repertoire of social interactions they hoped to capture.

Happily, on the last day of filming, their perseverance was finally rewarded. As they headed out of the harbour a group of mothers and calves swam up and started playing near the boat, leaping out of the water, bowriding and energetically slapping their tail fins in the sea.

But the highlight of the shoot was yet to come – as the family group continued to play, three young males swam up to investigate the commotion.

Male dolphins are known to cooperate to hassle females into mating with them, and it soon became clear this was the youngsters' goal. As they came nearer the team were able to capture courtship behaviour never seen before in Welsh seas, with the males spinning, nudging and jostling the females just a few metres from the boat. Eventually one male struck lucky and drew a female aside to mate. A few minutes later the group broke up and the dolphins disappeared as rapidly as they had arrived. The crew returned to New Quay, the remarkable footage they obtained revealing new insights into the social lives of these extraordinary animals.

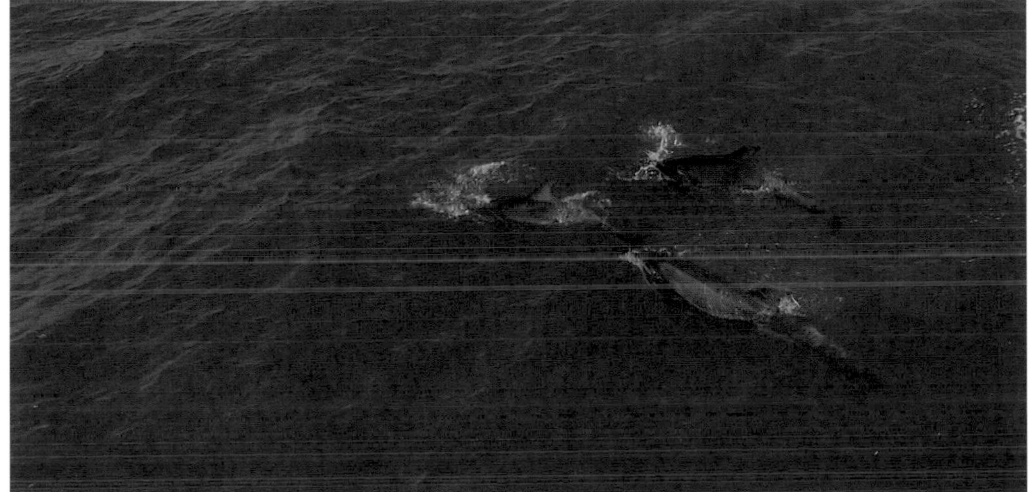

As well as capturing some magical footage of a family group of dolphins playing together, the *Celtic Deep* team filmed rarely seen dolphin courtship and mating behaviour.

Shallows:
Mythical Sea Kingdom

Lying where the warm Gulf Stream meets the cold-water currents descending from the Arctic, the shallow seas surrounding Wales contain some of the richest and most varied marine wildlife in the whole of the UK.

Hidden beneath the sea just off the coast of Wales lies a complex mosaic of submarine habitats as diverse as any to be found on land, ranging from dramatic rocky reefs to dense kelp forests and from lush seagrass meadows to vast sandy deserts. This magical otherworld is populated with creatures straight from Welsh legend, transforming colour, shape and even sex. Here, everything is in flux and nothing is what it seems.

Like their terrestrial counterparts, the inhabitants of the shallow seas are all ultimately dependent on the sun. In spring the lengthening days and warming seas trigger the growth of blooms of plankton, providing a lavish non-stop buffet which sustains a rich, interconnected web of life. Beneath rugged sea cliffs, steep underwater rockfaces and overhangs are cloaked in dazzlingly coloured anemones and corals.

The strongest tidal currents occur where the sea is funnelled into narrow channels, and one of the best examples of this occurs off the Pembrokeshire coast. On a racing tide, as the sea gets squeezed between the mainland and the saw-toothed rocks known as the Bitches, the water can reach speeds of up to 18 knots. Stirring up nutrients from the seabed, this attracts shoals of fish, which in turn draw in harbour porpoises, dolphins and seals.

Along more sheltered stretches of the coastline, sand and gravel cover the seabed, providing a home for a wide variety of burrowing animals, including one of the shallow seas' supreme masters of disguise, the Atlantic bobtail squid.

The most biodiverse habitats occur where sunlight penetrates through to the seabed: underwater kelp forests provide food and shelter for a multitude of different marine invertebrates, each with their own weird and wonderful adaptations, and are important hunting grounds for fierce hunters such as the small spotted catshark. In a sheltered bay on the north coast of the Llŷn Peninsula, a rare seagrass meadow provides a vital nursery for fish, including a true-life dragon of the Welsh seas, the snake pipefish.

Humans have not always been kind to the inhabitants of the shallow seas – overfishing, insensitive coastal development and pollution have all inflicted huge damage. But one aspect of human activity has proved immensely beneficial. Scattered across the Welsh seafloor, thousands of long-drowned shipwrecks act as artificial reefs for a wealth of wildlife – from the gender-fluid cuckoo wrasse to one of the largest predators of the shallow seas, the giant conger eel.

The abundance of different habitats provides food and shelter for all kinds of strange creatures.

Arctic Terns: Following the Sun

Late spring in Cemlyn Bay on the remote north coast of Anglesey – the far edge of Wales. Here, on a narrow finger of land, the sunlight-powered explosion of life within the shallow seas connects to the skies above, attracting the most extraordinary of all sun seekers, the Arctic tern.

Effortless flyers, these small, elegant, black-capped seabirds make the longest migrations of any birds on earth, constantly following the sun to stay in perpetual summer. Every spring, they migrate between the Antarctic and Arctic circle – a round-trip journey of up to 35,000 km. Weighing just over 100 g, they're so light that ocean breezes can carry them for great distances without them having to flap their wings, and they sleep and eat on the wing.

Cemlyn Bay is one of the terns' most northerly stop-off points in the UK. Between April and August, they nest in noisy colonies on a couple of small islands within a shallow lagoon, sharing their coastal sanctuary with two other regular summer visitors, their close relatives the sandwich and common tern. The terns are here for one reason – the blooms of microscopic plankton in the surrounding seas have attracted large shoals of silver-coloured sandeels, a vital food source for their newly hatched chicks.

Arctic terns are surface feeders – on spotting their prey the parent birds hover over the sea like a hummingbird, then plunge their feet or beaks in the water to skim fish from just below the surface. Returning to the lagoon, they are in constant communication with their hungry chicks. With nothing but a shallow scrape in the shingle to protect them, the tiny youngsters are highly vulnerable to predators – their parents are renowned for the ferocity of their sharp-beaked attacks on any intruder foolish enough to get too close.

By the end of summer, the young are strong enough to fly and the Arctic terns are ready to return south. They'll be back next year, back to the shallow seas of Wales, back in search of the sun.

Left: Cemlyn Bay on the north coast of Anglesey provides a safe haven for a rare breeding colony of Arctic terns. In summer adult birds can be seen fishing in the shallow coastal waters, on the hunt for sandeels to feed their growing chicks.

Bottom: Once the chicks are strong enough, the whole colony flies back to the Antarctic, a round trip of around 30,000 km.

Arctic terns defend their chicks fiercely and will dive-bomb any intruders.

FACT FILE
ARCTIC TERNS

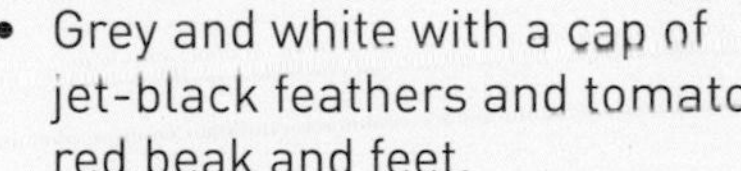

- Grey and white with a cap of jet-black feathers and tomato red beak and feet.
- Can be distinguished from common terns by their very long tail feathers and lack of a black splash at the tip of their beak.
- Length: 33–35 cm.
- Arctic terns see more daylight than any other animal on earth.
- Each year, from April to August, they come to breed on the remote north coast of Anglesey.
- Large shoals of sandeels are critical for the survival of the young chicks.
- Elsewhere in the UK, a steep decline in sandeels has led to the collapse of many tern colonies.

Kelp Forests

Stretching out beyond the rocky shoreline lies a magical realm – a submarine kelp forest. Reliant on direct sunlight to photosynthesise and grow, kelp forests are only found in waters down to 30 m and thrive in Wales' cool, nutrient-rich shallow seas. The dense stands of kelp provide food, homes and shelter for an astonishing variety of marine invertebrates and fish – including the largest wrasse in Welsh coastal waters, the striking-looking ballan wrasse – and are a favourite hunting ground for top predators, such as the Atlantic grey seal and the bottlenose dolphin.

Just like a terrestrial woodland, the kelp forest has a distinctive vertical structure. The top layer, or canopy, is formed by the leafy fronds of the largest seaweeds, the brown kelps. These need lots of light to grow and so are only found in the clearest waters. Growing up to four metres high and coated in mats of tiny hydroids and bryozoans, the kelps provide the main structural framework of these wild underwater woods.

Beneath the canopy the kelps' long trunk-like stems, or 'stipes', form the forest's understorey. This supports the greatest range of marine animals. As well as creating a sanctuary for large shoals of fish and crustaceans, such as the spiny spider crab, the stipes also provide a mooring for filter feeding sponges and anemones as well as other smaller seaweeds that reach towards the life-giving sun. At the base of the kelps, strong, branching holdfasts anchor the stipes to the rocky seabed, their numerous cracks and crevices providing a refuge for around 250 different species of invertebrates, including brittlestars, crabs and sea snails.

Lower down, on the seabed, is the forest floor – tangled thickets of smaller red seaweeds which need less light to grow.

This shrubby undergrowth is home to some of the strangest inhabitants of the forest, tiny sea slugs, with around 100 different species living in the shallow seas of Wales. These bizarre-looking creatures are renowned for the advanced weaponry with which they defend themselves from predators and their highly colourful sex lives.

In addition to supporting an astonishing range of biodiversity, kelp forests also absorb large amounts of carbon and help cushion the coastline from destructive storm-born waves.

Hidden beneath the water, the kelps' thick trunk-like stems, or 'stipes', support the leafy canopy, creating large areas of tranquil water.

Top: A ballan wrasse swims through the kelp forest. Up to 60 cm long and fiercely territorial, ballan wrasse use their thick lips and strong teeth to feed on small crustaceans living amongst the seaweeds.

As well as providing homes for thousands of species of animal, the kelps provide a mooring for other smaller seaweeds which reach up towards the sun.

Right: Eat or be eaten – the shrubby red seaweeds are home to many species of carnivorous sea slugs, which feed on tiny anemone like animals called hydroids. The stubby horns protruding from their backs are tipped with stinging cells absorbed from their prey, providing a highly effective poisonous shield against predators.

A sea hare feeding on a bed of shrubby red seaweed. These strange-looking molluscs take the colour of whatever they eat – red on red seaweed and green on green.

Sea Hares: Live Fast, Die Young

Sea hares get their name from the large ear-like tentacles projecting from their heads. Members of the highly diverse seaslug family, these bizarre-looking creatures are most frequently spotted in summer, when they congregate in long mating chains on shrubby beds of seaweed.

Like all molluscs, sea hares have a strong muscular foot which they use to glide around their leafy surroundings. At the front end of the foot lies the head, consisting of a pair of fleshy feeding tentacles behind which are two tiny eyes. Lacking a hard protective external shell, they defend themselves from attack by conjuring up a dark purple ink to foil approaching predators.

Unlike their close relatives the nudibranchs, sea hares are strict vegetarians – at the back of their large feeding tentacles is a toothy tongue, or 'radula', which they use to tear off the tough fronds of seaweed. As well as food, the seaweed also provides a valuable source of camouflage, with the sea hares taking on the colour of whatever they eat, turning green in green seaweed and maroon-red in red.

However, it's not their remarkable colour co-ordination that marks out these unusual forest dwellers but their steamy sex lives. For many marine animals it can be hard to find a mate in the vast and unpredictable sea. Sea hares are further handicapped as they only live for a year. With so little time to spare, they have come up with a particularly ingenious solution to partnering up.

Sea hares are hermaphrodites – they are both male and female. While hermaphroditism is not uncommon in the shallow seas, sea hares exploit the potential benefits to their ultimate limit. Once one pair hooks up, they release chemical signals into the water which attract other sea hares. Linking up with each other to form a long chain, each individual acts as male at the front end and female behind, ensuring that it both donates and receives sperm. Soon, long spaghetti-like strands of eggs can be seen coiling around the bed of seaweed, each one containing several million eggs. Lasting up to three days, this prolonged loveathon is impressive to witness.

However, there's a steep price to pay for this lascivious lifestyle. Once mating is over, the sea hares retreat into the shrubby seaweed and die shortly afterwards. But living fast and dying young has its advantages – it's thought that mass mating increases the likelihood of having genetically strong offspring, which may explain the sea hares' abundance in the highly competitive world of the kelp forest.

Above: In summer, long mating chains of sea hares surrounded by wispy pink strands containing millions of eggs can be seen on the forest floor.

FACT FILE
SEA HARE

- Size: 7–20 cm long.
- Reddish maroon, green or brown, depending on the colour of the seaweed they feed on.
- Upper head tentacles resemble hares' ears – giving them their common name.
- When disturbed or threatened by predators they eject a dark-coloured ink.
- Breed from early May to October.
- Hermaphrodites.
- Mate in long chains with each individual acting as both male and female.
- Produce pink spaghetti-like strands containing millions of eggs.

Small Spotted Catshark

Easily recognisable by their long torpedo shape, slanting green eyes and small dark spots, catsharks are a frequent visitor to the kelp forest. Although no more than a metre long, their extraordinarily acute senses make them one of the most fearsome predators of the shallow seas.

Like all sharks, the catshark's skeleton is made of light, flexible cartilage rather than bone. Along with their streamlined shape, this built-in suppleness enables them to weave effortlessly through the environment.

Catsharks are formidable nocturnal hunters, using their excellent eyesight, sense of smell and electroreception to track down their prey.

FACT FILE
CATSHARK

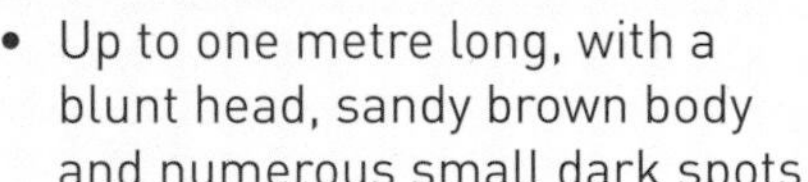

- Up to one metre long, with a blunt head, sandy brown body and numerous small dark spots.
- Live in shallow waters from 4-50 m in depth.
- Instead of scales, their skin is covered in 'denticles' – tiny backward-facing teeth.
- Nocturnal hunters, they rely on their excellent eyesight, sense of smell and electroreception to detect their prey.
- Catsharks' eggs are popularly known as mermaids' purses.

A catshark rests on top of the forest canopy.

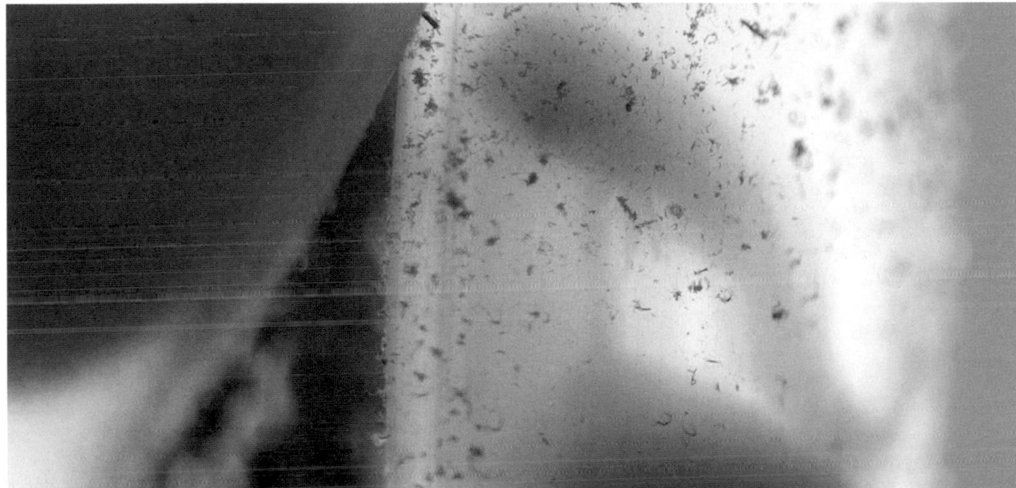

During the day they are relatively inactive, and can often be seen resting on top of the forest canopy, but at night it's a different matter. With their keen sense of smell and excellent eyesight, catsharks are highly effective nocturnal hunters, capable of putting on incredible turns of speed. Just like cats, they have a layer of reflective cells behind their retina, enabling them to see in dark or murky waters. Their favoured prey are crabs, whelks and bottom-living fish such as dabs and gobies. Most animals living on the seabed know to stay well hidden when the catshark is on the prowl. However, even then they can't be sure of avoiding detection, for the catshark has a supernatural power – electroreception. The smallest of electrical currents generated by a tiny movement or a beating heart can be enough to give the game away. This remarkable sixth sense enables catsharks to detect and root out shellfish and worms even when hiding beneath the seabed.

The complex architecture of the kelp forest provides the ideal nursery for the catshark. After mating, the pregnant female seeks out a sheltered spot within the forest understorey. Laying her leathery rectangular eggs two at a time, she then swims around her chosen mooring. By snagging the long tendrils protruding from the corners of the eggs, this ensures they won't drift away during the long gestation period. Inside each egg a large yolk sac sustains the growth of the tiny embryo developing inside. After eight to nine months, a tiny baby catshark hatches out and swims off into the underwater forest.

Left: Female catsharks lay their eggs deep within the the kelp forest. The developing embryo feeds on the yolk sac inside the egg, hatching out some eight to nine months later.

An inquisitive tompot blenny peers out from the rockface. It will live in the same crevice for several years.

Submarine Reefs and Cliffs

Constantly battered by wind and waves, the rocky cliffs of Wales are home to relatively few plants and animals, but beneath the waves it's the exact opposite – as the rockface plunges downwards, its jagged slopes are transformed into exotic gardens of anemones and corals, forming a dazzling kaleidoscope of shapes and colours.

This rocky submarine landscape is the Welsh equivalent of the Great Barrier Reef. Lying where the warm waters of the Gulf Stream meet the cool nutrient-rich seas from the north, it's home to an astonishing array of living creatures, all ultimately dependent on the same food source – the millions of microscopic plankton proliferating in the sunlit seas. These tiny animals and plants are the vagrants and gypsies of the ocean, drifting around on the tidal currents that sweep the coastline. For any creature that can hold on tight to the underwater landscape, they provide plentiful food on tap.

The underwater reefs lying within Skomer Marine Conservation Zone off the coast of Pembrokeshire are home to a particularly rich diversity of marine life, including 132 sponge species (from the latest Skomer sponge diversity survey 2019) and over 40 species of anemone, from the rare scarlet and gold star coral to exquisite jewel anemones. With the brightly coloured tips of their tentacles, anemones may look like flowers, but these strange, soft-bodied creatures are actually animals – their tentacles are armed with thousands of tiny stinging cells which they use to trap and kill their prey.

One of Skomer's rarest underwater residents is the flamboyant sea fan. Looking rather like old-fashioned quill pens, sea fans belong to the coral family and are usually only found further south in the Mediterranean. Each fan consists of a slender branching skeleton; covered with fleshy tissue, it's made up of hundreds of tiny animals called polyps, each one armed with stinging tentacles. The fans usually grow out from the rockface at right angles to the prevailing currents, enabling each individual animal to trap much more plankton than if it lived alone. Incredibly slow growing, some of the fans are thought to be over 100 years old.

Elsewhere, critically endangered spiny lobsters and others like the tompot blenny find food and shelter amongst the rocks.

Jewel anemone.

Dahlia anemone.

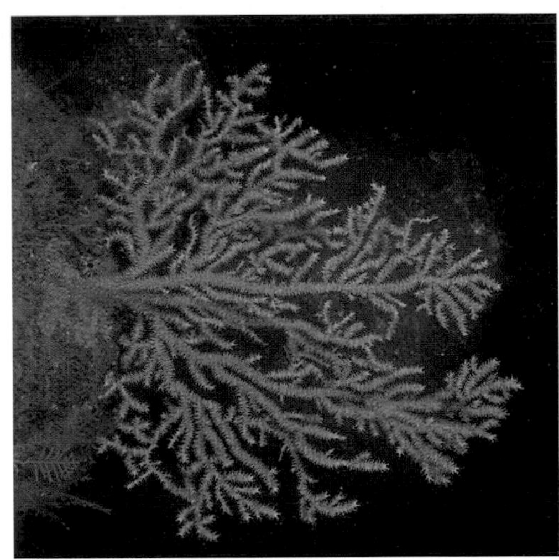

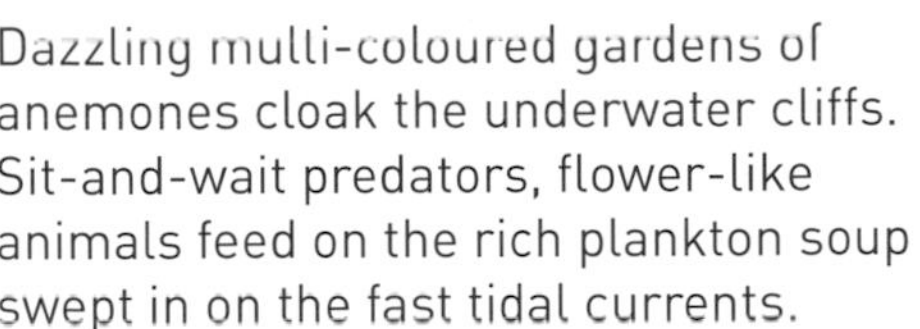

Dazzling multi-coloured gardens of anemones cloak the underwater cliffs. Sit-and-wait predators, flower-like animals feed on the rich plankton soup swept in on the fast tidal currents.

Above: A rare sea fan grows out from the underwater rockface.

The rare scarlet and gold star coral is more usually found in the warmer seas of the Mediterranean.

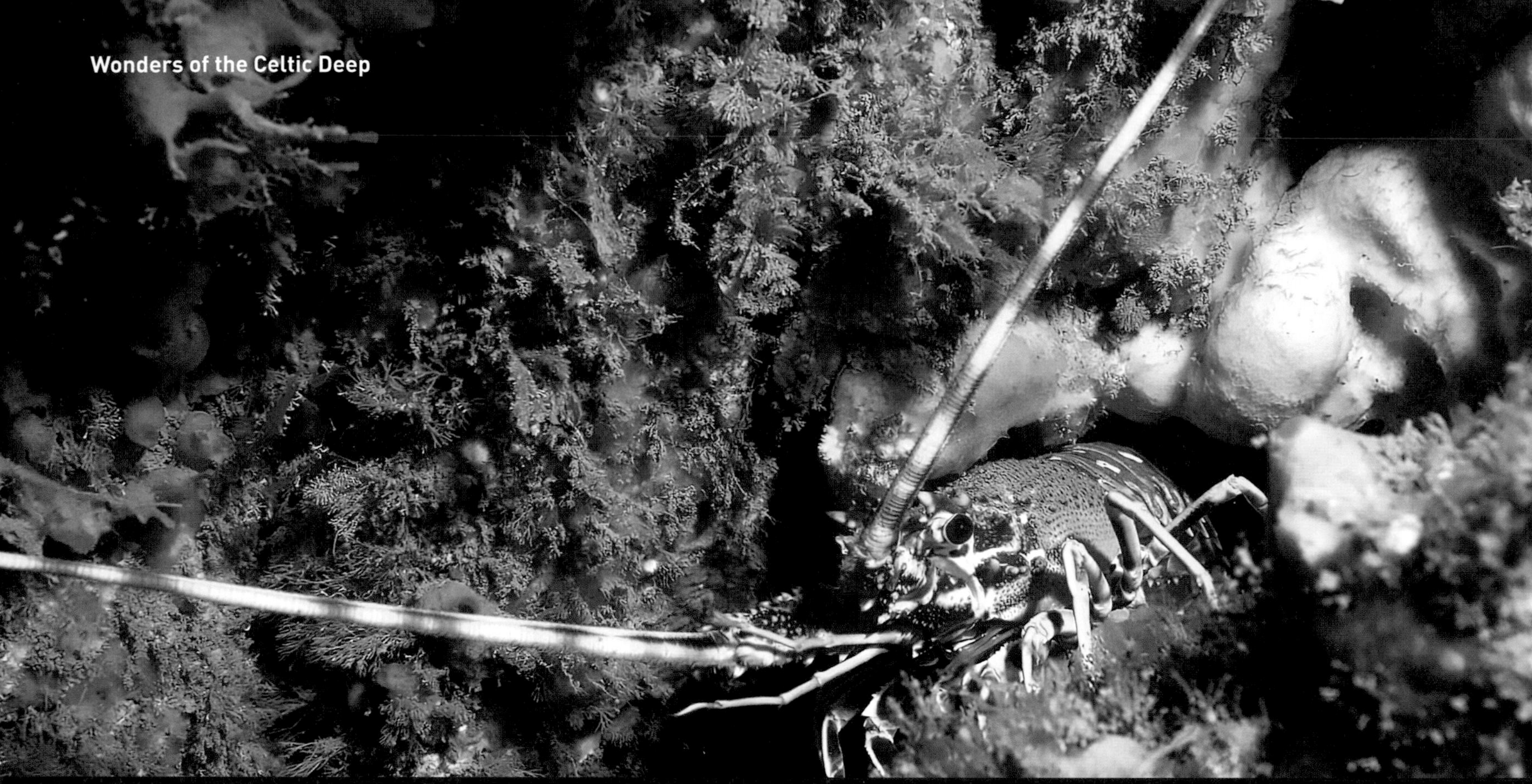

Spiny Lobster

Up to half a metre long and protected by a thick, spiky carapace, spiny lobsters are one of the apex predators and scavengers of the underwater reef. Hunting forays usually take place under cover of darkness, and while they prey on a wide array of shellfish, mussels are a particular favourite. Unlike the common lobster, which can easily crush through shells with its large pincers, spiny lobsters only have two small hook-like claws to dismember their prey. Delicately prising apart the two shells of the mussel, they use their claws rather like a knife and fork to extract the soft flesh inside. A series of elaborate appendages then act as a conveyor belt to waft the shreds of flesh into their mouths. A single mussel may take up to half an hour to consume.

During the breeding season the females can become very noisy; rubbing their antennae together, they create a creaking noise to attract a mate. Females produce hundreds of thousands of eggs, which they carry on their abdomen for several months. After hatching out, the larvae spend the next six months floating amongst the plankton. Despite their tiny size, they too are carnivorous, feeding on fish larvae and other microscopic animals.

Spiny lobsters used to be widespread along the Welsh coast until overfishing caused a crash in the population, and by the mid-1980s they vanished from many areas where they had previously been common. After 40 years, low numbers are slowly beginning to be found again at some sites along the coasts of Wales.

Above: Spiny lobsters get their name from the sharp spines which project from their thick carapace and antennae. Living in small groups, they spend much of the time hiding away in rocky crevices.

Right: Mussels are one of the spiny lobster's favourite foods – they use their small hook-like claws to prise apart the shells and extract the flesh.

FACT FILE
SPINY LOBSTER

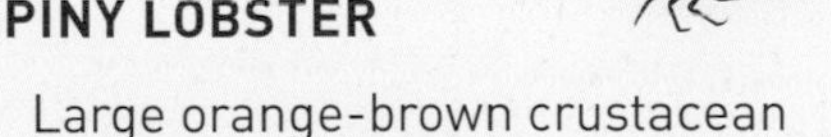

- Large orange-brown crustacean up to half a metre long.
- Spiky body armour, small hook-like front claws and two long spiny antenna.
- Live in small groups in rocky crevices in shallow waters, 20–70 m deep.
- Hunters and scavengers, feeding on shellfish, echinoderms, shrimps and worms.
- Known as 'creakers' in west Wales due to the rasping sounds the female makes to attract a mate.

Underwater Deserts

Beyond the sheltered bays and beaches of the Welsh coastline, the seafloor slopes downwards and transforms into a vast seabed of sand and shingle stretching out towards the open sea. This submarine desert may appear empty and barren, but a closer look reveals a remarkable wealth of life.

With no obvious places to hide, the inhabitants of these underwater deserts are superb masters of disguise. They need to be, for the shallow seas of Wales are patrolled by fierce predators, including the small spotted catshark, with its supernatural sixth sense.

Body shape, colour and pattern are all used to evade detection. Bottom-dwelling flatfish, such as the sandy-coloured brill, hide in plain sight, blending into the seabed so well that it's virtually impossible to see. Other animals take a completely different approach – hermit crabs decorate their shells with a garden of multi-coloured seaweeds and translucent fanworms to disguise their presence, while the Atlantic bobtail squid burrows beneath the sand, leaving only its huge eyes peering above the seabed.

Beyond sheltered beaches the sand extends out to create a vast submarine desert.

The need to hunt while being hunted has led to an evolutionary arms race between predator and prey, most notably in the age-old battle between the king scallop and its arch enemy, the spiny starfish. Other animals, such as the giant spiny spider crab, rely on formidable body armour to withstand attack.

For most animals it's not possible to hide away all the time – in the breeding season they need to advertise their presence to attract a mate. One of the most unusual methods employed in the search for a partner is that of the male painted goby, which emits repeated low-frequency drum beats to entice potential mates across the seabed.

Brill.

Fan worm.

Hermit crab.

Top: With nowhere to hide from predators, most animals living on the seabed have developed highly sophisticated camouflage. The brill's flat shape and sandy colouration makes it near impossible to spot on the seabed.

Bottom: This hermit crab has decorated its shell with seaweeds and a translucent fan worm to escape detection.

Painted Gobies

These small, rather insignificant-looking fish have evolved remarkably sophisticated courtship behaviour which cleverly exploits the fact that sound travels much faster underwater than in air.

During the breeding season the males actively seek out abandoned shells to set up home. On locating suitable living quarters the male defends his shell fiercely, raising his ice-blue coloured fins to scare off any rivals and producing long and repeated pulses of drumbeats – punctuated by the occasional thump.

Having seen off the competition, the next challenge is to attract a mate. Sound travels 4.3 times faster in water than in air, making his drum beats a highly effective way to advertise his presence. Once a female swims into view, the male attempts to grab her attention by energetically quivering his body and swimming elaborate figures of eight. However, it's not the quality of his body moves but rather the calibre of his drumming which determines the outcome – recent research has revealed that the males producing the longest drum riffs are most likely to entice females into their shells to mate.

There is an impeccable logic to the female's mating strategy – as with many fish, it's the male who looks after the eggs: those producing the longest pulses of drums are likely to be the strongest and fittest of her potential suitors and so most likely to successfully fend off predators. Once the female has entered the shell and laid her eggs she swiftly departs, leaving the male to guard them until they hatch.

Below: Painted gobies live on sandy and gravelly seabeds and can be distinguished from other gobies by the lines of black spots running along their dorsal fins.

In the breeding season males develop bright blue dorsal fins. After setting up home in an empty shell they defend it fiercely, generating low-frequency drum beats to fend off their rivals and attract a mate.

FACT FILE

PAINTED GOBIES

- Small, sandy-coloured, bottom-dwelling fish, around 5 cm long.
- Distinguishable from other gobies by the one or two lines of black spots running along their dorsal fins.
- Fish have more ways of creating sound than any other vertebrates.
- Male gobies use visual displays and pulsing drum beats to defend their territory and attract a mate to their shell home.
- After mating, the female lays small pear-shaped eggs, with the male guarding them until they hatch.

Spiny Spider Crabs: Warriors of the Shallow Seas

Encased in formidable body armour, spiny spider crabs are the great warriors of the seafloor. Up to 20 cm wide, their dome-shaped carapace is studded with thick spines on top with longer spines projecting around the rim. As they march across the desert terrain they wouldn't look out of place in a *Star Wars* movie.

Spiny spider crabs spend the winter away from the coast in the deeper ocean, but in early summer, as the water temperatures rise, they migrate in their tens of thousands into the warm, sheltered seas off the Welsh coastline. Here they roam the seabed, scavenging for seaweeds, mussels and starfish. Male spider crabs are notoriously combative. In one particularly memorable shoot off the Llŷn Peninsula, the *Celtic Deep* diving team came across a pair of males battling it out for dominance, their huge pincers crashing down on each other until eventually one of them backed off in ignominious retreat.

Despite their impressive suit of armour, spiny spider crabs are vulnerable to sharp-toothed predators, such as seals and dolphins. As additional protection they actively camouflage themselves, collecting seaweeds, barnacles and sponges to stick on and add to their already bizarre appearance.

Below: The largest crab in Welsh seas and equipped with a thick, spiky suit of armour and mace-like pincers, spider crabs are the great warriors of the shallow seas.

The Gathering Army

By late summer the spiny spider crabs have finally reached the end of their long journey across the seabed. Beneath the waves, a remarkable spectacle is about to play out. Congregating on the seafloor, more and more crabs start to pile up, creating huge mounds of up to a hundred individuals. The weary warriors have gathered for one reason – to shed their hard outershells. While their thick armour provides excellent protection, shedding their carapace is the only way they can grow. But until they develop a hard new shell the soft-bodied creatures inside are easy prey – piling up on top of each other provides the best protection while their brand-new suit of armour develops.

By early autumn thousands of discarded crab shells can be seen washing up on Welsh shores. Meanwhile, beneath the sea, the hidden army of spider crabs heads back to deeper waters. While their lives out in the open sea remain largely mysterious, it's thought that mating occurs soon after they've moulted and their shells are still comparatively soft. The crabs then hunker down until the following spring before commencing another epic journey back to the shallow seas.

Above: In late summer spider crabs gather in huge numbers on the seabed to moult. Piling on top of each other provides protection from predators until their new carapace starts to harden.

Bottom left: Shedding their hard carapace is the only way young crabs can grow. Once they have their new suit of armour they head back to deeper waters to spend the winter.

FACT FILE
SPIDER CRABS

- Grow up to 21 cm – the largest crab in Welsh waters.
- Orange dome-shaped carapace covered in spines and with long, spidery legs.
- Camouflage themselves by adding algae, barnacles and sponges to their shells.
- Live in deep water up to 120 m in winter, migrating into coastal waters in early summer.
- Scavengers, they feed on seaweeds, mussels and starfish.
- Lifespan: five to eight years.
- Moult in large mass gatherings in late summer before heading back to deeper waters.

Atlantic Bobtail Squid

Courtship is usually initiated by the female. Mating can last up to 80 minutes.

It may be no bigger than a golf ball but the Atlantic bobtail squid is a ferocious hunter. With its huge eyes it can easily spot its favourite food – tiny mysid shrimps – capturing them in lightning strikes using its two long retractable tentacles. But the shallow seas are a dangerous world and the hunter can quickly become the hunted. Like its much larger relatives, squids and octopuses, the bobtail lacks a protective outer shell, making it an easy meal for the many predators patrolling the seabed.

During the day it spends most of the time hiding just beneath the surface of the seabed. Covering itself with sand and gravel to leave only its eyes exposed, it remains on constant watch for both predators and prey. Pale cream with brown and black splodges, it blends in perfectly with the sandy seafloor.

At night the bobtail completely switches lifestyle. Shooting out from under the sand it uses its pair of semi-circular fins to hover above the seabed, on the lookout for its prey. Like most cephalopods bobtails are masters of disguise. Beneath its skin an underlying layer of iridiophores reflect and polarise light; the resulting green, blue and silver iridescence matches perfectly with the surrounding sea.

FACT FILE
ATLANTIC BOBTAIL

- Small cup-shaped cuttlefish, with eight arms and two retractable tentacles which it uses to capture its prey.
- Around 5 cm in diameter.
- During the day it buries itself just beneath the seafloor.
- At night it becomes an active hunter, using its huge eyes to spot its main food, tiny shrimps.
- Highly sophisticated camouflage – can abruptly change skin colour by expanding and contracting tiny sacs of pigment called chromatophores.
- Female initiates mating, which lasts for 60–80 minutes.
- Lifespan of around a year.

On top of this their skin also contains tiny sacs of pigment called chromatophores – by opening and contracting these they can change colour within a split second, making them practically impossible to see.

Moonlit nights offer the best hunting opportunities, for bobtails have one further conjuring trick up their sleeve – a specialised organ called a photophore contains bioluminescent bacteria that emit light at the same wavelength as the moon. This effectively masks the bobtail's silhouette, creating a fabulous cloak of invisibility to equal anything encountered in Welsh myth and legend.

When it comes to courtship, the female leads the way. On spotting a male, she hovers close by until he notices her. With a life span of no more than a year, Atlantic bobtails have to make the most of these chance encounters. Mating is a prolonged affair – lasting well over an hour – with the male using a specially adapted arm to transfer packets of sperm into the female's body cavity. After mating the female lays up to 100 eggs, which she then covers carefully with sand and gravel before vanishing into the night.

Top: During the day the bobtail hides under the seabed with just its huge eyes showing.

Middle and bottom: At night the bobtail becomes an active hunter. By expanding and contracting tiny sacs of pigment beneath its skin, it can change colour within a split second.

Old Battles: King Scallop vs Spiny Starfish

Amongst the fierce battles for survival playing out across the desert, the epic encounters between the king scallop and its arch enemy, the spiny starfish, is one of the oldest and most intriguing.

Easily recognisable by their ribbed, fan-shaped shells, king scallops spend the day sitting snugly in shallow hollows on the seabed, filtering out the rich plankton soup via their thick fringe of tiny tentacles. But this sedentary lifestyle leaves them highly vulnerable to one of the most voracious predators of the seabed – the spiny starfish. Evolving around 450 million years ago, starfish are superb stealth hunters, detecting their prey through sight and – more importantly – smell. Beneath each arm are hundreds of tube feet, at the tips of which are tiny suckers that can sense minute chemical traces wafting from their prey.

But the king scallop also evolved a long time ago and – over millions of years – has developed a highly effective defence strategy to avoid being eaten. Firstly, it has remarkable eyesight. Lining the edge of the soft mantle is a ring of 200 tiny metallic-blue eyes, each of which contains a tiny mirror that focuses light onto the retina, providing a highly effective early warning system for the feeding scallop.

To avoid detection the spiny starfish attacks from the rear, slowly and silently inching closer on its tube feet. This is where the scallop's next defensive weapon comes into play, for this seemingly immobile creature is capable of surprising turns of speed. As the starfish gets within touching distance it rapidly opens and closes its shells and shoots off backwards. Looking rather like a pair of flying castanets, it can swim several metres across the seabed. Although it can only sustain these energetic bursts for a few seconds, it's usually enough to escape the clutches of its age-old enemy.

Above: A king scallop watches out for predators with its ring of metallic-blue eyes. On detecting an approaching starfish it claps its shells together and jets off.

FACT FILE
KING SCALLOP

- Large bivalve mollusc, up to 21 cm in diameter.
- Evolved around 270 million years ago.
- Ridged, pale orange, fan-shaped shells. The upper one is flat, while the lower is bowl-shaped.
- Sits in shallow hollows on the seabed, filter feeding on plankton.
- 200 metallic-blue eyes keep watch for predators.
- Can live in protected areas for more than 20 years.
- Commercial scallop dredging has devastated the surrounding marine wildlife in large areas of the shallow seas.

FACT FILE
SPINY STARFISH

- Large pale grey or blue echinoderm – up to the size of a dustbin lid.
- Each of their five arms has three rows of large spines, with purple eyespots at the tip.
- Voracious stealth predators, hunting bivalve molluscs, crustaceans and other starfish.
- Detect their prey by sight and smell.
- Move across the seabed on their hundreds of tube feet.

On detecting its prey the spiny starfish launches a stealth attack, creeping up from behind on its hundreds of tiny tube feet.

Seagrass Meadows

Porthdinllaen harbour, on the north coast of Wales. The clear, sheltered waters provide the perfect conditions for a rare seagrass meadow.

The Llŷn Peninsula, north Wales. Nestling beneath a slender ribbon of cliffs curving into Caernarfon Bay lies the picture-perfect fishing village of Porthdinllaen. Hidden under the crystal clear waters of its sheltered harbour is one of the rarest and most important marine habitats in the world – a seagrass meadow. The size of 46 rugby fields, it's the largest in Wales. As luxurious as the best-kept garden lawn, the seagrass meadow is made up of eelgrass, one of only a few species of flowering plants that can survive fully submerged in seawater. In June, its long, narrow leaves explode with tiny seedpods, which scatter through the sea to create new meadows.

Like the kelp forest, seagrass needs sunlight to grow, so is only found in shallow waters. Beneath the waves thousands of creatures can be found sheltering within this unique ecosystem. As well as providing food and sanctuary for lobsters, spider crabs and squid, these lush green pastures harbour one of the most important fish nurseries along the entire Welsh coast.

In summer the meadow is alive with fish of all shapes and sizes – from sinuous eels to corkwing wrasse and polka-dotted plaice. Here, amongst the gently swaying grasses, parental care is often provided by the male of the species. The slender ribbons of grass create safe mooring for the lovingly constructed nest of stay-at-home dad, the 15-spined sea stickleback, and its sheltered waters provide the pregnant male pipefish with plentiful food until his 200-strong brood hatches out.

As well as supporting a wealth of marine wildlife, seagrass meadows bring many environmental benefits. Unlike kelp, seagrass has proper roots. By helping to stabilise the soft, sandy seabed, it provides a vital buffer against destructive storm waves. The meadows are also an important tool in the fight against climate change, soaking up even more carbon dioxide per square metre than tropical rainforests. Sadly, along much of the coast, damage from boats, agricultural runoff and insensitive coastal development has wiped out these unsung wonder plants of the shallow seas. So rare is the Porthdinllaen seagrass meadow that a major conservation project is underway to collect and transfer millions of seeds 150 miles south to the Dale Peninsula in Pembrokeshire – the biggest seagrass restoration project ever undertaken in the UK.

A plaice swims through the seagrass meadow.

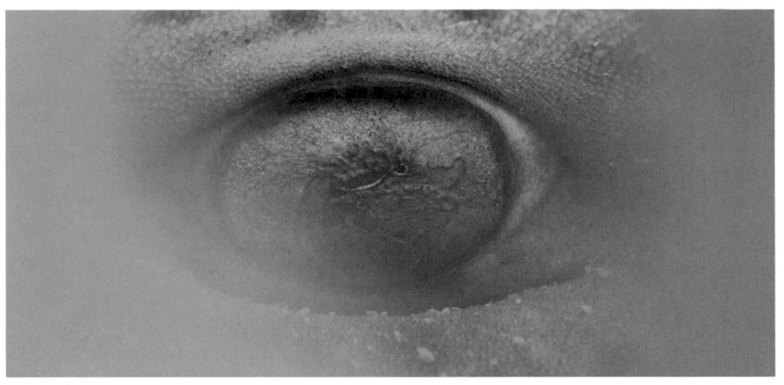

A bull huss shark keeps a watchful eye on its neighbours.

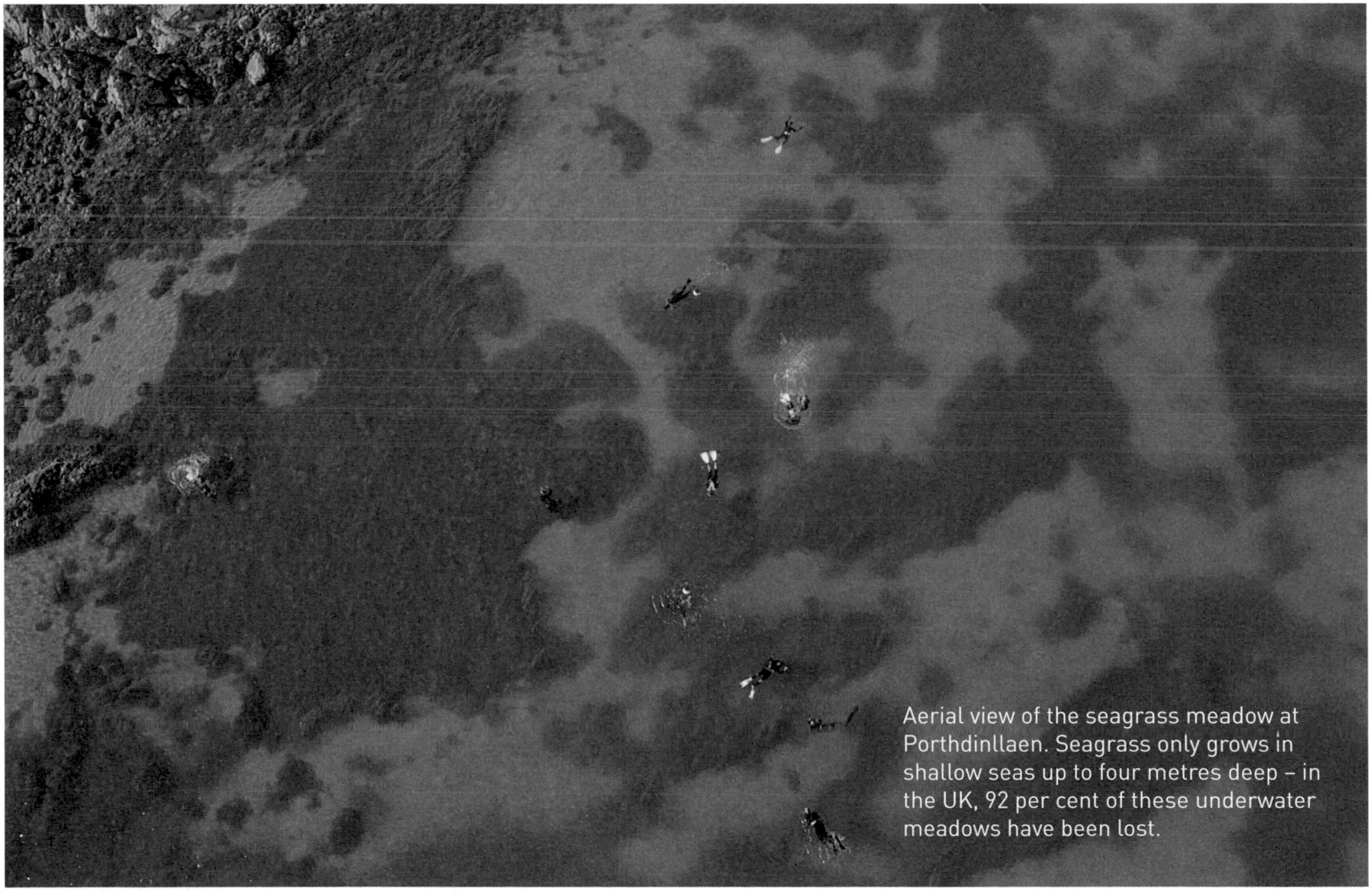

Aerial view of the seagrass meadow at Porthdinllaen. Seagrass only grows in shallow seas up to four metres deep – in the UK, 92 per cent of these underwater meadows have been lost.

Pregnant Dads and Stay-at-Home Fathers: Pipefish and Sea Sticklebacks

One of the most striking-looking residents of the seagrass meadow is the snake pipefish. Up to 60 cm in length and with a long tubular snout, they are closely related to their much curvier cousin, the seahorse. These real-life dragons of the Welsh seas have excellent camouflage, deliberately swaying their stripey orange-brown bodies in rhythm with the rippling seagrass.

Pipefish spend the day weaving gracefully through the meadow, on the hunt for their favourite foods – tiny mysid shrimps – which they suck up into their straw-like mouths. Spotting their highly mobile prey in this vast three-dimensional habitat could be a real challenge, so pipefish have come up with an ingenious solution. As well as being able to rotate their eyes up, down and sideways, each eye can move independently from the other, enabling the pipefish to home in on its prey while continuing to look around for its next meal.

When it comes to raising offspring, pipefish definitely push the boundaries. After a lengthy courtship dance with her prospective mate, the female transfers her eggs into a small hollow in his tummy. After fertilising the eggs, the male carries them around inside him until, several weeks later, he gives birth to tiny pipefish babies. Being a pregnant dad has its benefits – males can be assured they are looking after their own offspring whilst giving them the best protection possible.

Above: A snake pipefish swims through the seagrass on the hunt for tiny shrimps.

FACT FILE
SNAKE PIPEFISH

- Up to 60 cm long, the largest pipefish in Welsh waters.
- Orange-brown in colour, with a red band extending back from the tip of its long snout.
- Each eye can move independently, so the pipefish can train one eye on its target while looking for the next meal.
- Males carry eggs in a small hollow in their tummy before giving birth to live young.

Like the snake pipefish, the male 15-spine sea stickleback is solely responsible for parental care. In early summer, as the shallow seas begin to warm, the male starts building a nest. Collecting wispy strands of seaweed, he weaves the delicate fronds around the seagrass before carefully glueing them together with a sticky gum secreted from his kidneys.

Males invest considerable effort into constructing a comfy nest; however, it's not the quality of his self-build that is most likely to impress prospective mates but how high it is off the sea floor. Large predators, including lobsters and bull huss sharks, roam the meadow and fish eggs are an easy meal.

On entering the nest, the female lays up to 200 eggs, after which she swims off leaving the male in sole charge of the crèche. As well as fending off predators, parental duties include fanning the eggs with his tail to keep them well oxygenated. Around two to four weeks later, the eggs hatch out and the young sticklebacks swim off to take their chances in the big wide world of the meadow.

Above: A male sea stickleback builds his nest from tiny strands of sargassum seaweed. The higher the nest above the seabed, the more likely he is to secure a mate.

FACT FILE
SEA STICKLEBACK

- Olive-brown in colour and up to 22 cm long.
- Front dorsal fin consists of 15 small spines.
- Male builds a nest 4–5 cm in size high in the seagrass.
- Females will stake the male's nest before deciding whether to mate.
- Lifespan: one to two years.
- Both parents die soon after breeding – the female after laying her eggs and the male after hatching.

SS *Gwynfaen*: Submarine City

Along the Welsh coast the lethal mix of ferocious storms, powerful tidal currents and treacherous rocky reefs have claimed the lives of many seafarers. Lying just beyond the rugged coastline, ghostly wrecks litter the seafloor. But for many marine animals, these long-drowned vessels are one of the best places to set up home, providing abundant food and shelter in the unpredictable seas.

A mile offshore from Porthdinllaen lies the remains of the SS *Gwynfaen*, a 300-tonne steamship which came to grief in rough seas over a hundred years ago. Today its decaying carcass has become a bustling submarine city, full of life. Fleshy mounds of soft coral – the aptly named Dead Man's Fingers – coat the steel girders that mark the submarine city limits and huge sea urchins the size of saucers mow up and down the rusting hull, grazing on the thick crust of barnacles.

Weaving in and out of the twisted pipework are shoals of gold sinny, poor cod, bib, pollack and cuckoo wrasse, which would otherwise steer well clear of the flat, sandy seafloor that surrounds the wreck. But, like any large metropolis, the *Gwynfaen* is not without its dangers. Hiding out in the engine room is one of the apex predators of the shallow seas, the giant conger eel, and the densely inhabited city provides it with an amazing range of top-quality eateries. To protect themselves from their troublesome neighbour, the fish rely on a highly sensitive internal sonar system. Extending along their flanks, a line of tiny sensors – the so-called 'lateral line' – detect minute vibrations in the water, enabling them to keep track of each other's movements even at night. If any individual senses danger, the entire shoal darts away into the safety of the *Gwynfaen's* many hidden nooks and crannies.

Dead Man's Fingers – a type of soft coral – filter feed on the plentiful plankton wafting through the wreck.

The *Gwynfaen* acts as an artificial reef for large shoals of fish, such as these silvery coloured bib. Elsewhere, sea urchins graze on the barnacles coating its outer girders.

Dead Man's Fingers.

Shoaling provides safety in numbers. Here a shoal of silvery bib swims through the wreck.

FACT FILE
SHIPWRECKS

- Shipwrecks act as artificial reefs, creating a valuable source of food and shelter for marine wildlife.
- SS *Gwynfaen* was heading from the Llŷn Peninsula to Birkenhead when she hit rocks off Porthdinllaen headland.
- Today the 300-tonne steamship is home to a dazzling array of animals, including conger eels, common lobsters, bib, cuckoo wrasse and the Bloody Henry starfish.

Sea urchins graze on barnacles coating the ribs of the old steamship.

A female conger eel peers out from its hideaway in the *Gwynfaen*'s engine room.

Conger Eel and Common Lobster

Deep in the engine room of the *Gwynfaen* lives one of the true giants of the shallow seas – a female conger eel. Up to three metres long, these fierce, opportunistic predators are the largest eels in the Celtic Seas. Shipwrecks are highly desirable pieces of coastal real estate, providing abundant nooks and crannies to hide in and a constant supply of fresh food. Unlike the sociable bib, congers are real loners – they spend the day resting, with just their large heads peering out from their carefully selected hideyhole.

Hunting takes place almost exclusively at night. Emerging from her refuge she reveals a dark grey body with a white underside and long, continuous fin running along her back. Congers track down their quarry by sight and smell. Armed with a set of sharp, densely packed teeth and powerful jaws, they have the capacity to devour most of their neighbours. Prey ranges from fish – including smaller congers – to squid, large crabs and even lobsters. But despite the conger's fearsome reputation, small groups of common shrimps are often seen congregating nearby to feed on the ample leftovers.

Congers live almost their entire lives as immature fish. Between 5 to 15 years old they embark on a one-way trip hundreds of miles into the deep Atlantic to spawn. Here the female lays several million eggs, dying soon afterwards. The larvae float around in the ocean currents; those that survive develop into small eels before commencing their mammoth swim back to the safety of the Celtic Seas.

FACT FILE
CONGER EEL

- Largest eels in Welsh seas, growing up to three metres long, with females much larger than males.
- Common in and around wrecks.
- Grey-blue to grey-black with a paler underside.
- Smooth skin with no scales.
- Formidable nocturnal predators, feeding on fish, lobsters, squid and crabs.
- At 5–15 years they migrate into the deep Atlantic to spawn.
- Lifespan: Up to 20 years.

Living next door to the female conger is another longstanding resident of the *Gwynfaen*, a common lobster. With its dark blue and yellow carapace, long red antennae and broad tail fan, it's arguably the wreck's most magnificent beast.

Lobsters feed on a wide range of prey, including fish, crabs, clams, mussels and sea urchins. Around half a metre long, their front legs bear two powerful pincers, one much larger than the other. While the larger claw is used to crush through shells, the smaller one carves up the flesh.

Despite their massive body armour, lobsters are vulnerable to many predators, including seals, octopus and, of course, humans, which is why the abundant nooks and crannies found in shipwrecks make them such an attractive refuge. Like other crustaceans, they can only grow by shedding their hard external skeleton, with moulting happening every one to two years.

The common lobster is one of the longest-living creatures in the world – some animals are believed to be over 100 years old. Theoretically – if they manage to evade predators – lobsters can live forever. It's thought that their longevity may be due to the presence of telomerase, an enzyme that repairs the DNA caps at the ends of chromosomes. Unlike other animals, this persists into adulthood. Ironically, one of the most likely causes of death is another lobster – fiercely territorial, lobsters will fight to the death to fend off challengers.

Above: The common lobster's massive claws are capable of crushing through bone.

FACT FILE

COMMON LOBSTER

- Large crustacean, average length around half a metre, but can grow up to a metre.
- Dark blue and yellow carapace, with long red antennae.
- Front legs bear two pincers, one usually larger than the other.
- Crushing claw exerts 45 kilos per square inch of pressure.
- Usually move around by walking but can also swim rapidly backwards.
- Use chemical signals to form dominance hierarchies and find mates.
- Capable of living up to 100 years.
- Overfishing in the 1980s led to an 80 per cent drop in numbers but they are now making a slow comeback.

Cuckoo Wrasse

The brightly coloured cuckoo wrasse may look as if it belongs in the Caribbean, but these striking-looking fish are widespread off the Welsh coast. In addition to their dazzling tropical costumes, they are renowned for their remarkable gender-fluid sex lives.

Cuckoo wrasse live in small shoals, made up of a dominant male and several females, and shipwrecks provide the perfect sanctuary. As they swim through the wreck they feed on the bountiful supply of small fish, molluscs and crustaceans, their powerful teeth and jaws easily crushing through tough shells. Males and females can be easily distinguished by their colouring – while the male has a brilliant blue head with bright orange and blue marked flanks and fins, female wrasse are a rich coral pink, with a long fringe of black and white blotches.

Like most marine creatures, finding a mate can be a real challenge, so these sexual sorcerers have come up with a particularly inventive solution. Their sexual sorcery makes them one of the most extraordinary shape-shifters to be found in the entire magical kingdom. All cuckoo wrasse begin life as females, but if the leader of the group dies then the largest female transforms into a fully functional male. As if preparing for their new identity, transitioning females can often be seen swimming amongst the group – their colours a mixture of male and female. There are clear benefits to swapping sex – as well as ensuring there is always at least one male to mate with, it pays to start out as a female and then turn into a male when you are larger and better equipped to fend off any rivals.

Above: A female cuckoo wrasse swims through the wreck. All cuckoo wrasse start out as female, but as they mature some start transitioning into males.

Spawning takes place in late spring and early summer. After building a nest of seaweed, the male performs an elaborate swimming dance to woo his prospective mate. Once the eggs have been laid, he fiercely defends the nest. After a few weeks, the eggs hatch out and the larvae float away, eventually developing into little female wrasse and repeating their extraordinary life cycle.

Above: Male cuckoo wrasse can be distinguished from females by their blue heads and the bright blue and orange markings along their flank and tail.

FACT FILE
CUCKOO WRASSE

- Territorial fish, common on shipwrecks and rocky reefs lying 20–80 m deep.
- Males up to 40 cm long, females usually smaller, around 30 cm.
- Males have a bright blue head and orange and blue markings along their flank and tail.
- Females are coral pink with a black and white blotched dorsal fin.
- All start out as females but some undergo a dramatic colour change and start transitioning into males.
- Live in small groups headed by a dominant male. Once he dies, a transitioning female takes his place and becomes a fully functional male.
- Maximum lifespan: 20 years.

Shallows: Diaries
Filming the SS *Gwynfaen*

Filming the marine wildlife living on the twisted wreck of the *Gwynfaen* was always going to be one of the most challenging shoots of the series.

Filming the SS *Gwynfaen* Day 1

As the sun sets over Caernarfon Bay, on board the *Highlander* Director of Photography Rob Taylor makes final checks ahead of the night shoot.

Diving off the coast of Wales is always unpredictable: the ever-changing weather has a huge impact both on underwater visibility and on the marine wildlife – no matter how well prepared you are, you never quite know what you're going to find.

Shipwrecks bring their own additional complications – the window for filming is limited to the period of slack water in between tides, leaving just one hour to capture the amazing animal communities living on board. On top of this, most marine creatures are more active at night, with all the potential hazards that brings when filming beneath the sea.

On the plus side, the *Celtic Deep* filming and diving team had a huge amount of experience filming in difficult sea conditions. Director of Photography Rob Taylor, marine biologist Rohan Holt, diving supervisors Rich Stevenson and Martin Sampson and underwater cameraman Jake Davies worked closely with researcher Summer Kiernan and location manager Jet Moore to develop a dive plan that allowed for all eventualities.

Right: The diving team make their way down the shot line and into the wreck of the *Gwynfaen*, 20 m beneath the surface.

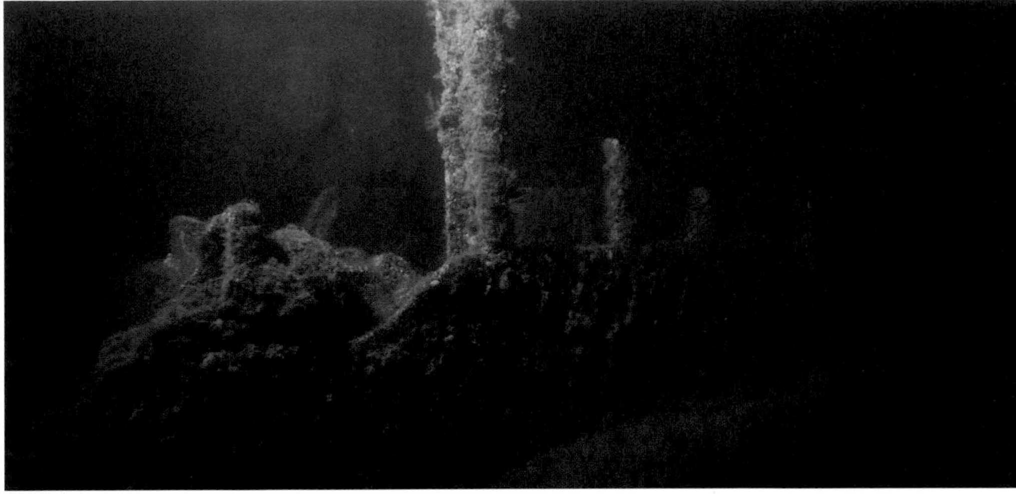

Even so, as the team set off into Caernarfon Bay on a fine summer evening, there was no guarantee of success.

Top of the wish list was one of the alpha predators of Welsh seas – the conger eel. Two to three metres long, these real-life giants do most of their hunting at night. With shipwrecks one of their favourite haunts, the team had high hopes of capturing a conger cruising through the rusting hull in search of supper. But, as they made their way down the shot line, it became immediately obvious it was going to be a tough shoot. The underwater visibility was appalling and a strong current was ripping through the wreck. Guided by Jake and Rohan, Rob and Rich carefully made their way towards the engine room, where conger eels had previously been spotted. Initially, it looked as if luck was on their side – from deep within the twisted pipes of the boiler, a female conger eel's beady brown eyes gazed out at them. Unfortunately, she was clearly in no mood to embark on any hunting sorties that night, not even when tempted with some tasty scraps of mackerel. One hour later the team emerged from the murky black water into the cool night air. When I asked Rob how it had gone, he glumly said it was one of the worst dives he'd ever done.

Left: Unfortunately, the female conger eel refused to leave the boiler.

Day 2

Having rested and regrouped, and with the fine weather continuing, the team decided to return to the *Gwynfaen* in daylight, hoping that the underwater conditions had improved. Before heading off to the wreck, they took some time out to film a pod of playful bottlenose dolphins bow riding behind the boat. The happy-go-lucky dolphins definitely brought better fortune – when the dive team returned underwater the conditions were crystal clear, revealing an astonishing profusion of life – large shoals of bib and pollock, giant sea urchins, a magnificent common lobster and the dazzlingly coloured, gender-fluid cuckoo wrasse, all there to be captured on camera. In spite of the limited time underwater, Rob wove his usual magic, transforming the *Gwynfaen* shoot into one of the best dives of the summer.

The day shoot proved much more successful, with the crew filming a spectacular array of marine wildlife both above and below water.

Into the Deep:
Exploring the Open Seas

The *Celtic Deep* team set out to explore the habitat and wildlife of Wales' open seas, a place difficult to get to, inhospitable for much of the year and with which we are least familiar. Here, as on land, it is governed by the changing seasons and, luckily for us, the most productive time of year is also the most conducive for going to sea and filming.

The two seas of Wales, the Celtic Sea to the south and the Irish Sea wrapping round the west and north, are a rich marine and aquatic habitat of worldwide importance. Bordering three oceanic and climatic zones (North-East Atlantic, Arctic Boreal, Lusitanian) and retaining a coastline with one of the largest tidal ranges in the world, this unique aquatic environment has some of the most diverse marine wildlife in Europe. Its shores are epic, its coast is wild and its seas are rich enough to rival the tropics.

Several distinct types of seabed habitat provide ideal conditions for a variety of species, and the open ocean above grants perfect conditions for marine wildlife such as dolphins, whales, sharks, jellyfish and plenty more less well-known but equally fascinating inhabitants – many of them making impressive sea journeys in order to feed and breed here – and all of them exhibiting fascinating displays of natural behaviour.

The mission for the *Celtic Deep* team was to film as much of Wales' open ocean wildlife as they could during the busy summer and early autumn months, when food is at its most plentiful. But filming out in these treacherous waters would not just prove to be challenging but near impossible.

Surprisingly little is known about many of the species that travel these waters and coupled with the changing climate and 2020, a year when few scientists were able to go to sea, up-to-date knowledge was thin on the ground. The crew had to rely on a mixture of old data, tip-offs and educated guesswork to film this episode.

Once a sighting has been made, the practical challenges of filming open ocean pelagic species are not to be underestimated. Boats are noisy, meaning the wildlife can hear you coming from many miles away, and with the visibility underwater never more than 12 m – only half the length of a fin whale – it's never far to escape the lens. Under the guidance of marine biologists and a filming license, the crew also had to adhere to strict rules in order to film the wildlife with minimal disturbance.

Safely getting a film crew into the water can then only be done in the most perfect of conditions – a calm, almost flat sea and no wind. Days like this are hard to come by at any time of year in the seas around Wales. Luckily, the *Celtic Deep* team had a couple of rare days when the elements came together and the seas of Wales put on a show.

The Smalls Lighthouse, 20 miles (32 km) off the coast of Pembrokeshire. It is the final sight of land before entering the Celtic Sea.

Manx Shearwaters

Skomer Island in early spring, looking back towards the Pembrokeshire coastline in the distance.

Puffins, like Manx shearwaters, return in vast numbers to breed in the burrows of Skomer and Skokholm.

Early spring sees millions of transatlantic seabirds arriving on Welsh shores to breed.

The islands and unreachable cliffs which litter the Welsh coast are, for many of these birds, the only time they will be on land. They arrive in March before the water warms and the plankton bloom, needing to time the arrival of their chicks for when food is most available.

South Stack on Anglesey and Pembrokeshire's many islands come alive with sound as puffins, gannets, razorbills, fulmers, kittiwakes and guillemots vie for space on these rocky outcrops.

Skomer Island off the Pembrokeshire coast is the preferred choice of arguably Wales' most important bird, the Manx shearwater. Half the world's population, more than 600,000 birds, return here every year.

The islands of Skomer and neighbouring Skokholm are relics of Wales' violent, volcanic past. It touched the mainland of Wales in the last Ice Age, before rising sea levels cut it off. Their isolation, towering cliffs and, most importantly, a layer of rabbit-burrowed soil make it a haven for Manx shearwaters and other returning seabirds such as puffins, razorbills and storm petrels.

The shearwaters are here to breed, nesting deep in burrows on the clifftops, and choose islands like these for their rat-free status. The islands need to lack ground predators like rats, cats and foxes, which would happily dig a bird out of its burrow.

They may be safe in their underground nests, but there is no escape from aerial hunters. A Manx shearwater is a good-sized dinner for a gull, and there are lots here – herring, lesser and greater black-backed gulls all hunt these skies.

The patrolling gulls mean the Manx shearwater can only return to feed their chicks under the cover of darkness, spending the daylight hours far out to sea.

When the time comes for the chicks to fledge, they too must also wait for nightfall, when the predatory gulls have gone to roost. On a special night in September, the chicks emerge, head for the cliffs, and jump, tumbling 200ft to the sea below. The surviving fledglings then face a transatlantic flight across the Atlantic to South America for the winter, before hopefully returning to the islands of Wales to breed for years to come.

During the long days of summer the seas of the Celtic Deep are alive with Manx shearwaters, all with chicks to feed and their own bellies to fill before the long journey to South America.

Ungainly and impractical on land, these birds are masters of the air and amazing underwater. Using their wings as oars, they can swim down to depths of up to 55 m to hunt fish far below the surface.

Filming in amongst the chaos of the bait ball, the *Celtic Deep* camera team managed to record these elusive birds comfortably swimming around at depths of over 10 m.

Manx shearwaters on the surface above a bait ball. From there they dive down to pick off the fish below the surface.

FACT FILE
MANX SHEARWATERS

- Latin name: *Puffinus puffinus.* (Despite not being puffins, which are *Fratercula arctica*).
- Migrate over 10,000 km to South America in the winter.
- Wales' oldest living bird, found on Bardsey Island, was over 50 years of age.
- Can swim down to a depth of 55 m when hunting fish.

Gannets renew their bonds with a knocking of bills.

The Northern Gannet

With its two-metre wingspan, the northern gannet is the largest seabird in the north Atlantic.

Famous for its big appetite and skydiving hunting technique, the northern gannet can cover hundreds of miles when hunting, but like all seabirds needs to return to land to breed.

The tiny uninhabited island of Grassholm, eight miles off the coast of Pembrokeshire, is the nesting site of 10 per cent of the world's population. Every February, 78,000 gannets cram onto its ten hectares, turning the rock white with guano. Each pair lays a single egg on a nest made of seaweed and vegetation, although, sadly, the gannets of Grassholm are now in the habit of using discarded plastic and rope in their nests, resulting in entanglement and sometimes death.

Unusually for birds, gannets have no brood pouch, a bald patch under the chest for keeping eggs warm next to the skin, so they incubate the eggs with their feet which are filled with warming blood vessels.

When a chick hatches in early June, the parents have to work round the clock for 90 days to keep it fed. After three months of being stuffed with food, the chick is heavier than its parents, weighing in at a hefty 4 kg. When it is time to fledge 13 weeks after hatching, the chick doesn't fly the nest. Too heavy to take to the air, it has no choice but to jump and fall from the high cliffs of Grassholm into the Atlantic Ocean. Once on the water it is too fat and buoyant to dive for fish, and unable to either fly or dive, it swims until light enough and strong enough to take off from the water and dive for its first fish.

For the *Celtic Deep* series the focus was to film the gannets in action far out to sea. Highly sociable birds, they will often feed in large flocks, hunting herring, mackerel and sandeels. The sight of a large group of gannets diving in the distance provides a good clue as to where there are fish.

With excellent vision, their sharp eyes allow them to detect prey underwater from many metres above, despite the reflections and refractions of light interacting at the surface.

With the prey in sight, they perform dramatic plunge dives at up to 60 mph (100 kph), wings folded, often from as high as 40 m (130 ft), as high as eight double-decker buses. This trick enables them to catch fish at up to depths of 20 m.

They are incredibly well adapted to survive such high-impact hunting again and again, including being equipped with extremely strong neck muscles and a spongy plate of bone at the front of their skull. Gannets also have an extensive network of air sacs between their muscles and skin as additional cushioning against this impact.

FACT FILE
GANNETS

- 78,000 gannets fit onto Grassholm during nesting season.
- They dive from up to 40 m above the water to reach fish up to 20 m down.
- A chick is too heavy to fly or dive when it first heads to sea.
- They keep their eggs warm with their feet.

Common Dolphins off the Welsh Coast

Common dolphins (also referred to as short-beaked common dolphins) are one of the resident species to be found in the waters around Wales. A predominantly offshore dolphin, they can often be seen closer inshore when following schools of fish.

A gregarious and highly sociable animal, they are normally found in family groups known as pods.

Despite the fact that numbers of common dolphins off the UK's coasts have boomed in the last two decades, likely due to warming waters, it is unknown how many there actually are. The most recent survey in 2005–8 estimated the total UK population at 63,400.

The *Celtic Deep* crew didn't go a day without seeing large numbers, normally jostling for position off the boat's bow, and one particular day witnessed a group of many hundreds surrounding the boat and riding its wake – a truly spectacular sight, and not an unusual one.

Superpods of common dolphins are often observed off the coast of Wales, with the largest estimated to contain 2,000 individuals.

Superpod formation is thought to be based on food supply. If there is an abundance of food, a large group that hunts as one can take advantage of the opportunity, as individuals cooperate to panic and corral fish, making them easier to catch. Likewise, if there is a predator threat, a large group provides security.

Above: Large groups of common dolphins are a familiar sight off the Welsh coast, with superpods of thousands gathering on occasion.

These large groups are often seen travelling at speed. Sound travels 4.3 times faster in water than in the air and so if a pod locates a large school of fish, other groups within 10 miles or so will also hear about it and almost as quickly join the hunt – the chase becomes a race with themselves.

It's not known quite how such a large group communicates to hunt and stay together, but sound is thought to play an important role. Dolphins produce a whistle-type noise by vibrating tissues called 'phonic lips' in their nasal tooth cavities.

Some of the largest superpods contain thousands of individuals and are made up of sub-groups, smaller pods of 20 to 30 dolphins connected through relation or factors like age and sex. It is thought that when groups come together, individuals recognise each other from previous gatherings, using signature calls and individual names for each other.

A chance to socialise could also be a factor in bringing together groups of this size. Like humans, dolphins derive a sense of enjoyment from socialising in large groups. It is also believed that a lot of mating goes on during these gatherings, maximising the spread of genes by reproducing with individuals from different pods or families whose paths otherwise wouldn't cross.

FACT FILE
COMMON DOLPHIN

- Travel in large groups numbering between 10 and 50 animals, and occasionally hundreds, if not thousands.
- Estimated UK population of 63,400 and rising.
- Can reach speeds of up to 60 kph (or 37 mph) in short bursts.
- Grow up to 2.4 m long.

Plankton: The Unseen Hero

Spring comes late to the Celtic Deep – after a long winter the water takes time to warm, much longer than it takes the land, but when it does it triggers the start of a spring bloom.

Tiny 'bits' floating on the surface of the sea make the water look murky and 'dirty', but are in fact made up of billions of plant-like cells called phytoplankton and tiny animals called zooplankton.

Plankton are the unseen heroes of life on this planet, influencing almost every aspect of our lives, including the air we breathe.

The microscopic phytoplankton contain chlorophyll, with which they photosynthesize just like plants on land, drawing in carbon dioxide and releasing oxygen. They account for around 50 per cent of all photosynthesis on earth, and therefore 50 per cent of the oxygen we breathe.

Phytoplankton are also responsible for most of the transfer of carbon dioxide from the atmosphere to the ocean. Worldwide, this 'biological carbon pump' transfers about 10 gigatonnes of carbon from the atmosphere to the deep ocean each year. Their importance in storing away carbon dioxide in this way has made them a target for projects looking at controlling carbon dioxide in the atmosphere.

As diverse as plants on land, the ones featured in *Wonders of the Celtic Deep* are diatoms and are so small they had to be filmed under a microscope by plankton expert Dr. Richard Kirby. Diatoms are just one example of the many shapes and forms of phytoplankton found in Welsh waters.

As the season progresses, increased sunlight and warming water fuels the growth of phytoplankton, creating large plankton blooms and kick-starting the food chain in the Celtic Deep.

Below: Glass-like diatoms encased in a silica shell. Diatoms are found in every aquatic environment on Earth.

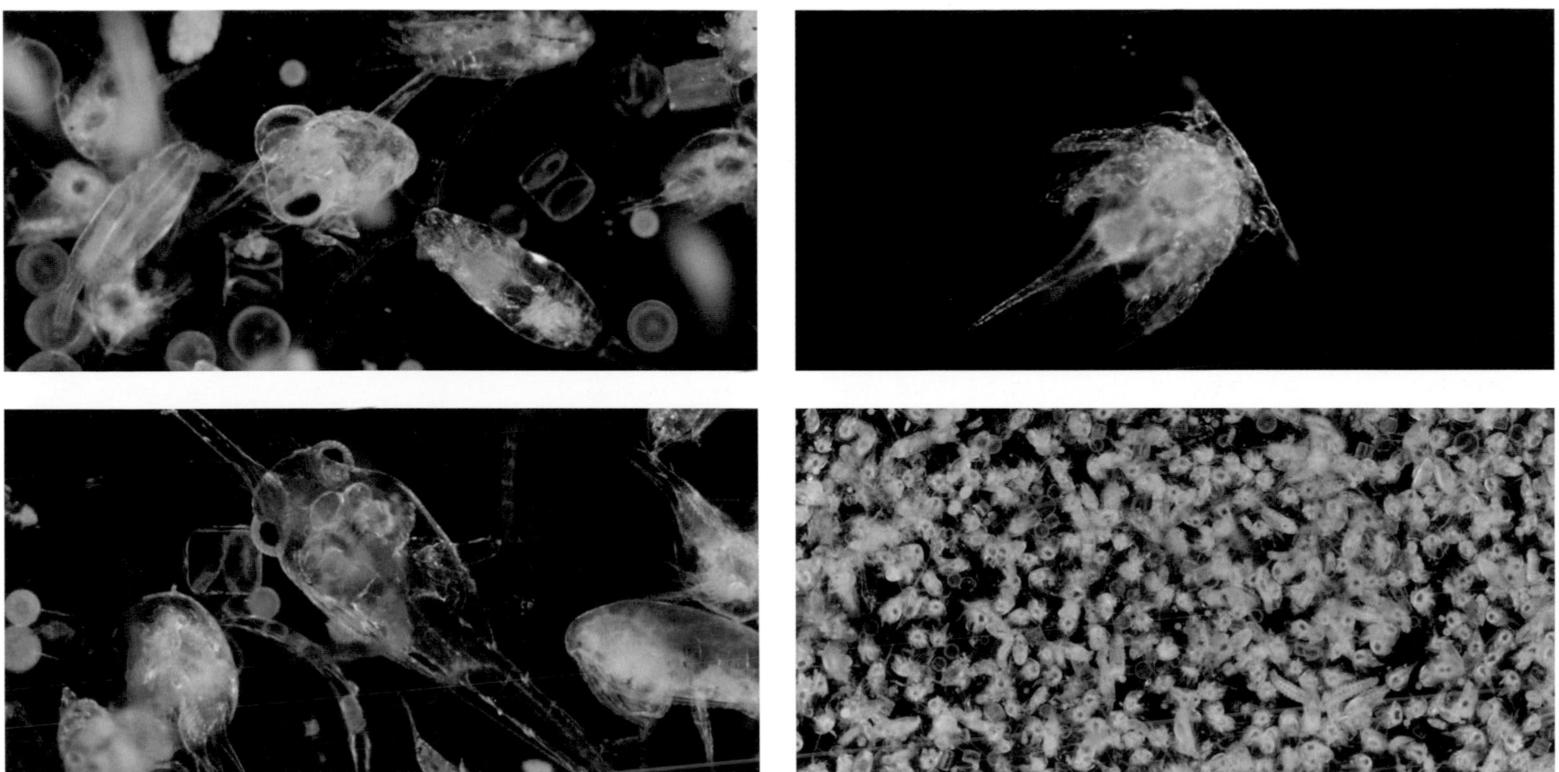

The plant-like phytoplankton are grazed on by tiny animals called zooplankton, which in turn are food for larger creatures like the basking shark and jellyfish.

The term zooplankton covers an astonishing array of microscopic and some not so microscopic organisms. Many of these open ocean drifters spend their whole lives near the ocean surface, feeding on phytoplankton and other zooplankton. For others, this is only one part of their life cycle.

Plankton are comprised of two main groups, permanent members of the plankton, called holoplankton (such as diatoms, radiolarians, dinoflagellates, foraminifera, amphipods, krill, copepods, salps, etc.), and temporary members (such as most larval forms of familiar seaside creatures like sea urchins, starfish, crabs, lobster, some marine snails, most fish), which are called meroplankton.

The coastline and rockpools of Wales, along with the seafloor of the Celtic Deep are littered with a cast of characters who spent their early life as meroplankton.

Above: The platonic forms of barnacles, crabs and other familiar seaside species can be found in the plankton of the open ocean.

A feeding basking shark can filter up to 2,000 cubic metres (2 million litres) of seawater per hour. To do this, the basking shark needs to cruise slowly through the water at no more than 5 kph (3 mph).

Basking Sharks, Follow the Sun and Plankton

Once the waters of the Celtic Deep start to warm, the larger pelagic open ocean species start to arrive.

Many of these open ocean wanderers pass through the Celtic Deep, spending the year moving to the most productive area of the Atlantic at any one time. For the most part this means following the sun and the blooming plankton it fuels.

Basking sharks are one of the first to arrive, turning up in April and May. After the whale shark, this is the largest fish in the world. They mature slowly and live for around 50 years.

Despite their size, they feed mostly on zooplankton, filtering up to 2,000 cubic metres (2 million litres) of seawater per hour through their gills.

Basking sharks have been listed as an endangered species for some time, but despite this very little is known about them, their numbers and their social lives. This is possibly because, for most of the year, the basking shark moves through the deep undetected.

A short window of opportunity to study this huge fish before it heads south in the autumn does come when they spend the summer inshore, feeding on the blooming plankton around the coast of Wales and the rest of the British Isles.

By winter they have migrated down the European continental shelf and moved to deeper water and to depths often exceeding 900 m (3000 ft).

It used to be thought that basking sharks were lonesome, independent predators, but recent research has turned that on its head and it is now believed that basking sharks may travel migratory routes in extended family groups, 'hanging out' with relatives, which could help with the learning of these routes. New research has also shown that they might not just be coming to the UK and Ireland to feed, but that they may also be using this time as a chance to socialise and breed.

Annual visiting behaviour off the UK's coasts has now been tracked using tagging with GPS locators and cameras, with adult basking sharks having been observed displaying odd swimming behaviours, possibly linked to courtship.

It appears that, typically, courting basking sharks will congregate in groups, swimming clockwise slowly in a tight circle. They have also been observed in small and tight aggregations near the seabed, fins touching but hardly swimming. This social swimming behaviour is similar to other shark species, but exactly how the interaction works between the basking sharks is still unclear.

Despite this behaviour now having been sighted and filmed a number of times, the actual mating has still yet to be witnessed and is believed to take place in deep water – far beyond the prying eyes of scientists and their drones.

In August 2020 a research drone from the Irish Basking Shark Group captured a rarely seen ritual – a group of circling basking sharks in what's thought to be courtship behaviour, although the mating has never been witnessed.

FACT FILE
BASKING SHARKS

- The second largest fish in the world, growing up to 12 m long.
- Females are pregnant for one to three years.
- A basking shark can live for around 50 years.
- IUCN lists them as an endangered species in the north-east Atlantic.
- Their liver accounts for around a quarter of their body weight.

Jellyfish in the Summer Months

A moon jellyfish, the most common jellyfish around the UK. Its four circular pink gonads are visible through its bell.

Jellyfish have drifted through the planet's oceans for more than 500 million years, despite lacking a brain, heart or blood. There are thousands of species around the world, but the seas around Wales are home to only six, most of which can be seen around the coast during the summer months as they follow in the wake of the plankton.

Jellyfish are carnivores, feeding on zooplankton, which get trapped in their tentacles and passed via cilia into a mouth, found under the middle of the bell, for digestion.

Jellyfish follow the plankton blooms arriving in the seas around Wales in late spring and summer, but being weak swimmers are at the mercy of the winds and currents – a big summer storm can result in millions of jellyfish being washed ashore and into coves and bays around the Welsh coast.

Often, these spectacles are made up of one particularly common jellyfish – the moon jelly. The moon jellyfish is easily identified by the four horseshoe-shaped gonads which can be seen in the middle of its translucent hand-sized bell. The bell and gonads may be translucent white, pink, blue, or purple, depending on the animal's diet. The tentacles that fringe the bell are lined with stinging cells, but luckily their sting is so mild that most people cannot feel it.

The lion's mane jellyfish, on the other hand, is one to avoid. It's easy to see where they get their name from – the long, flowing tentacles that surround the bell have the beauty and bite of a lion! Brown to reddish in colour, it has a thick mane of hundreds of long hair-like tentacles, the oldest of which are coloured dark red. Normally around 1-3 m in length (although in other parts of the world they

FACT FILE
JELLYFISH

- Jellyfish have no brain, heart, bones or eyes, although they can detect light.
- They have been in our oceans for over 500 million years, but as they have no hard parts, fossils are very rare.
- A group of jellyfish is called a 'smack' or 'bloom'.
- The lion's mane jellyfish is the longest recorded jellyfish in the world, with one recorded as reaching over 30 m long.

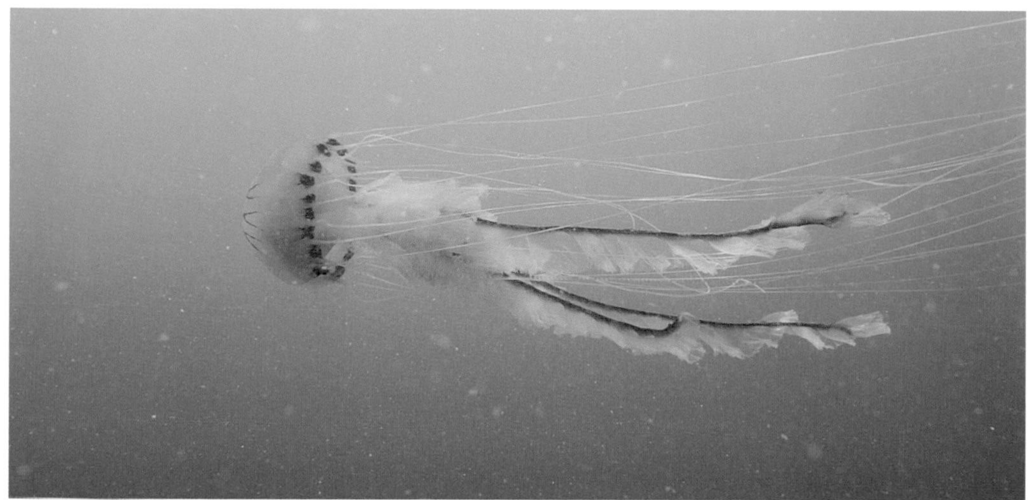

have been known to be record breakers), these tentacles are packed with stinging cells and are used to catch their favourite prey: fish and other smaller jellyfish.

Another jelly commonly found along the Welsh coast is the compass jellyfish. Named for the distinctive markings on its bell, this translucent yellowish-white jellyfish has brown compass-like markings around the fringe and on the top of the bell.

The largest jelly found in Welsh waters is the barrel jellyfish. As the name suggests, they can be the size of a barrel, with a translucent, milky-white, mushroom-shaped bell and a bunch of eight frilly tentacles below. They can often be found washed up on beaches in May and June.

Scientists have noticed that jellyfish numbers are rising across the world, possibly a sign of an imbalance in our oceans. Jellyfish thrive in water containing a lower than normal concentration of dissolved oxygen, with dissolved oxygen dropping in response to increased temperature or pollution. Jellyfish predators, such as leatherback turtles and ocean sunfish, cannot tolerate the same conditions, are subject to overfishing and climate change, and may die when they mistakenly eat floating plastic bags that resemble jellies. As a result, jellyfish numbers are expected to grow even further. But in the waters around Wales, following in the jellyfish's wake are some unusual-looking predators hoping for a chance to feast.

Top: A compass jellyfish and (middle and bottom) lion's mane jellyfish surrounded by a feast of plankton and small fish.

An ocean sunfish in the Celtic Sea. They spend much of their time hunting at depth before coming back to the surface to warm up.

Long-Distance Travellers

As the plankton bloom continues to fuel the food chain in the seas off Wales, more travellers arrive, some more associated with tropical waters.

Although not often seen, leatherback turtles are summer visitors to the Celtic Deep. They are cold-blooded but can elevate their body temperature to withstand our cold waters, travelling around 7000 km from the tropical waters where they nest to Wales' colder waters to feed. Despite their large size (up to 2 m long and around 700 kg) they feed on jellyfish and seem to like the local barrel jellyfish.

On their journey, leatherback turtles are at risk from accidental capture in fishing gear like longline fishing in offshore waters, or being entangled in ropes in coastal waters. The largest and heaviest turtle ever recorded was a 100-year-old leatherback turtle that washed up on Harlech beach, north Wales in 1988. Measuring almost 3 m (9 ft) in length and weighing 914 kg (2,016 pounds), it had sadly drowned after getting entangled in fishing gear. It now resides at Cardiff Museum and can be seen in the 'Man and Environment' gallery, next to a humpback whale.

A traveller more commonly seen and no less special is the ocean sunfish, or *Mola mola.* Weighing up to a tonne, the ocean sunfish is the largest bony fish in the world. Almost as tall as they are long (up to 4 m), they look like a large misshapen dinner plate and subsist mainly on calorie-low jellyfish. Famous for sunbathing at the surface and for being slow-moving, studies have shown that they only sunbathe for short periods each day, spending most of their time hunting jellyfish, fish and squid at great depths, where the water is cold, hence the wish to sunbathe when they come back to the surface.

The sun worshipping has another advantage: sea birds often congregate around them to pick off fish and crustaceans infesting them, and they will even flap when they are on the surface to attract birds to come and feast on their parasites.

Another of the sunfish's striking features are its huge eyes; giving it the equivalent in human terms of 20/100 vision, it's this which enables them to make out jellyfish in the water column. They also have eyelids, a rare thing in a fish, which means that sunfish can blink.

All in all, they are a wonderfully curious creature and brilliant visitor to come across in the seas off Wales.

FACT FILE

LONG-DISTANCE TRAVELLERS

- The largest and heaviest turtle ever recorded was a 100-year-old leatherback turtle that washed up on Harlech beach, North Wales in 1988. Measuring almost 3 m (9 ft) in length and weighing 914 kg (2,016 pounds).
- Leatherback turtles are named after the thick, leathery layer of skin that covers them. They don't have a hard, bony shell like other sea turtles.
- Leatherback turtles have downward-facing spines inside their throat that prevents any prey from getting out.
- Sunfish have eyelids, which means they can blink, a rare ability in a fish.
- The back fin which sunfish are born with never grows, resulting in their odd shape.

Below: Often more associated with tropical waters, both leatherback turtles and ocean sunfish are summer visitors to the Celtic and Irish Seas.

Atlantic Bluefin Tuna: Nature's Torpedoes

The 'ultimate fish', as well as an endangered species, Atlantic bluefin tuna are the largest of all bluefin tuna species and the largest fish in the mackerel family. Living up to 40 years, they grow to over 2 m (6 ft) in length and 600 kg (1300 lbs) in weight, undergoing remarkable transformations in size during their lifetime, from being nearly microscopic to one of the largest open ocean fish.

Bluefin tuna have been on human menus for centuries, but in the 1970s demand rocketed, particularly in Japan. As a result, prices soared and commercial fishing operations found new ways to locate and catch these sleek giants.

This consumer demand, particularly in Japanese sashimi markets, led to severe overfishing on both sides of the Atlantic.

Atlantic bluefin tuna are a slow-growing and long-lived species, leaving them particularly vulnerable to overfishing. For many years bluefin tuna appeared to be more or less absent from the waters around the British Isles and fishing and landing of them was, and still is, banned. In the last few years, however, the tuna appear to be back, with regular sightings in the Celtic Sea in particular, but this might not be the good news it appears to be.

Bluefin tuna are known to be highly migratory, and as transatlantic migrants their movements corresponding to spawning and food needs. Much of the specifics of their migration patterns remains unknown. They are loyal to their birthplace, returning to the breeding grounds from which they came.

Scientists believe that they are back visiting the Celtic Deep as a feeding ground following the increase in food availability during the late summer after having migrated from the Gulf of Mexico where they spawn, so this is less a rise in population numbers and more a result of movements in the population to the location where food is most plentiful; they remain classed as an endangered species on the International Union for Conservation of Nature's Red List of Threatened Species.

The Atlantic bluefin tuna is one of the most formidable predators to visit Welsh waters. Designed for speed, they are built like torpedoes, with retractable fins and their eyes set flush to their body. They hunt by sight and have the sharpest vision of any bony fish, seeking out schools of fish like herring, mackerel and even eels.

In short bursts, they can travel at a maximum speed of 40 kph (25 mph) and can dive to depths of more than 900 m (3000 ft), their versatility meaning little can escape them.

A truly remarkable fish, especially to witness them in action, the *Celtic Deep* crew were lucky to film tuna working bait balls of shoaling fish on many of their days at sea. Frantic splashing on the surface could be seen from a long way off, but it was only from the drone that the true scale of their activity was visible.

Below: An Atlantic bluefin tuna heading for a shoal of herring in the Celtic Sea.

Bottom: A drone's view of a tuna feeding frenzy.

FACT FILE

ATLANTIC BLUE FIN TUNA

- Lifespan of up to 40 years.
- One of the fastest fish in the world they can even retract their pectoral and dorsal fins to reduce drag.
- In short bursts, they can travel at a maximum speed of 40 kph (25 mph).
- Capable of diving to depths of more than 900 m (3000 ft).

Minke Whale

The minke whale is the smallest of the baleen whales and a resident species in the seas around Wales.

Like most of the world's whales, its population was decimated by whaling and the world population is now estimated at between 500,000 and 1 million individuals. Only a few UK population surveys have been carried out – one in 1994 estimated 8,500 individuals in the North Sea and Channel, but more recent reports suggest that the number is closer to 10,500 (including the Celtic Sea).

Minke whales are listed as 'Least Concern' on the IUCN Red List and like all cetaceans in the UK they are protected under the EU Habitats Directive (which persists despite Brexit).

Minke whales still face some whaling threats, however, and can become entangled in fishing nets and static fishing gear. Collisions with ships and anthropogenic-induced chemical/noise pollution also pose sizable risk.

As with their numbers, very little is known about their movements throughout the year – it is thought some migrate to warmer waters further south in winter, but others have been seen year-round off the Welsh coast.

Below: A distinctive white band on each flipper helps mark it out as a minke when size is hard to judge.

Calves are born between November and March after a 10-month gestation period, but we still don't know where their breeding grounds are and mother-calf pairs are a rare sight.

A short lactation period of four to five months means that the 2.5-metre calves are usually independent by the time they reach their summer feeding grounds.

Unlike some other whales, minkes are not cooperative hunters, preferring to go it alone in their hunt for fish, but, as with other baleen whales, they are filter feeders, with the 50 to 70 large pleats running along their throat allowing enormous volumes of water to be taken in during feeding. They gulp feed, primarily on krill and schooling fish like herring, sandeels, sprat, capelin and small cod, taking big mouthfuls of water.

Despite being mostly solitary, minke whales are notoriously inquisitive and are known to approach boats and spy hop to get a better view. If they are seen in small groups of two or three, they will often be strictly segregated by age and sex.

When filming, the crew saw a couple of minke whales but often briefly and sometimes not realising until they looked back at footage after the event. Minkes often turned up at a bait ball and, despite their size, managed to slip into the melee quietly and take their fill before disappearing again.

Above: Minke's have 50 to 70 large pleats running along their throat allowing enormous volumes of water to be taken in during feeding.

FACT FILE
MINKE WHALE

- Can be up to 10 m in length and weigh around 10 tonnes.
- Can live for up to 50 years.
- Minke whales' vocalisations can be as loud as jet planes taking off (up to 152 decibels).
- The smallest whale to be found in the Celtic Sea.

The Blue Shark

Blue sharks are elegant open ocean predators with a distinct pointed snout, saw-edged teeth and beautiful metallic dark blue backs which provide brilliant camouflage.

They exhibit 'countershading': a camouflage trick where the white grading on their undersides enables them to remain discrete from prey from all angles.

They have one of the largest geographic distributions of any shark, and historically – before being decimated by fisheries and the illegal fin trade – were one of the most (if not the most) abundant pelagic species in our oceans. Listed as 'Near Threatened' on the IUCN Red List, numbers have decreased by as much as 60–80 per cent in some areas.

An estimated 20 million blue sharks are still being lost from the world's oceans annually, with the illegal fin trade seeming the most wasteful. After death, the meat of the blue shark is rapidly contaminated by its own ammonia and so is not desirable; fins are often cut and the carcasses thrown back.

Blue sharks are highly migratory, using their large pectoral fins to ride long trans-oceanic currents to conserve energy as they make their seasonal commutes. They are known to travel thousands of miles in a single trip.

Most of the blue sharks found in the Celtic Sea are female, but no one knows why that is.

In the Atlantic, they appear to undertake a clockwise route, following the Gulf Stream to the UK from their eastern Atlantic breeding grounds and returning back via the Atlantic North Equatorial Current.

Blue sharks visit UK waters only in the warmer summer months, primarily for food but possibly also for mating, and are normally found more than 10 miles offshore. The waters around Pembrokeshire, which reaches out like an arm into the Celtic Deep, are home to large numbers of blue sharks, mostly mature females. No one knows quite why there's this abundance of females. Despite being extensively studied, there is little data on populations that reside in the north-east Atlantic.

Blue sharks grow relatively rapidly, maturing after four to six years. Courtship can be violent, with females often displaying bite marks after mating – their skin is three times thicker than males to avoid serious injury. Blue sharks are viviparous, giving birth to a brood of up to 50 live pups.

Feeding mainly on small fish like mackerel and herring as well as squid – although they are known to take seabirds and other sharks too – they are most active in the evening and at night, when they may move inshore.

Blue sharks will occasionally scavenge and have been spotted following trawlers, eating bycatch and gutted fish thrown overboard. Deoxygenation of the oceans is forcing them into shallower waters, making them more vulnerable to fishing pressures.

FACT FILE

BLUE SHARKS

- Largest caught in UK waters was 2.74 m (9 ft).
- Have a lifespan of around 20 years.
- Can dive as deep as 350 m (1,150 ft).
- Measuring up to 3.8 m (12.5 ft), females are bigger than males.

Horse Mussel Beds

The Irish sea around the Llŷn Peninsula in north Wales is the most southerly home of a rare northerly marine habitat – the horse mussel bed.

The horse mussel is a large cold-water mussel found in the northern Atlantic. Growing up to 20 cm long, these slow-growing mussels can live for up to 50 years and provide a home and structure for a whole community of seafloor dwellers.

Individual mussels on the seafloor bind together using byssus threads, hair-like 'beards', to form a habitat with many nooks and crannies. In doing this they become a type of biogenic reef – a reef made up of living organisms.

The structure they form then provides shelter for other species to hide from predators and act as important nursery and feeding grounds for diverse marine life, including commercially important species of fish and shellfish such as whiting, cod, queen scallops and common whelks.

Horse mussels don't reach maturity until they are over four years old, but make up for it by being extremely long-lived. This longevity means they are very slow to recover from damage, and much of the UK's horse mussel beds have now been lost due to disturbance, mostly from trawling.

The Welsh population is on the edge of its most southerly range and the Welsh horse mussel population is also thought to be under further stress from climate change and rising sea temperatures.

It is likely that the beds we now find around the Llŷn Peninsula are the last remnants of beds that were extensive following the last Ice Age. It's fair to say they used to be much more widespread in the Irish Sea, but the remaining horse mussel beds are considered important for its biodiversity and overall health. A healthy, mature bed is home to an astonishing range of species, all happily filling an ecological niche, and provides a solid foundation for creatures such as soft corals, tubeworms, barnacles and sea squirts as well as shelter for brittlestars, crustaceans, worms, molluscs and many other small animals.

Dead man's fingers litter the mussel bed, giving it an eerie look, and this soft coral is well named for looking like white fingers emerging from the sea floor. Each is actually a colony of individual small animals that share a gelatinous skeleton. The individual animals – or polyps – extend their bodies and tentacles out to feed on passing plankton – when feeding, the fingers appear furry.

From butterfly blennies to sea squirts and nudibranchs, the variety of life to be found living on a mature horse mussel bed is truly astonishing and just a snapshot of the interesting and varied life to be found on the floor of the Irish Sea.

Hundreds of species rely on the mussel beds for a home, from small fish and shrimp to clams.

FACT FILE

HORSE MUSSEL

- Latin name: *Modiolus modiolus.*
- Horse mussels can live up to 50 years.
- Grow up to 20 cm in length.
- Together they create a type of biogenic reef – a reef made up of living organisms.
- Under threat from: trawling, rising sea temperatures.

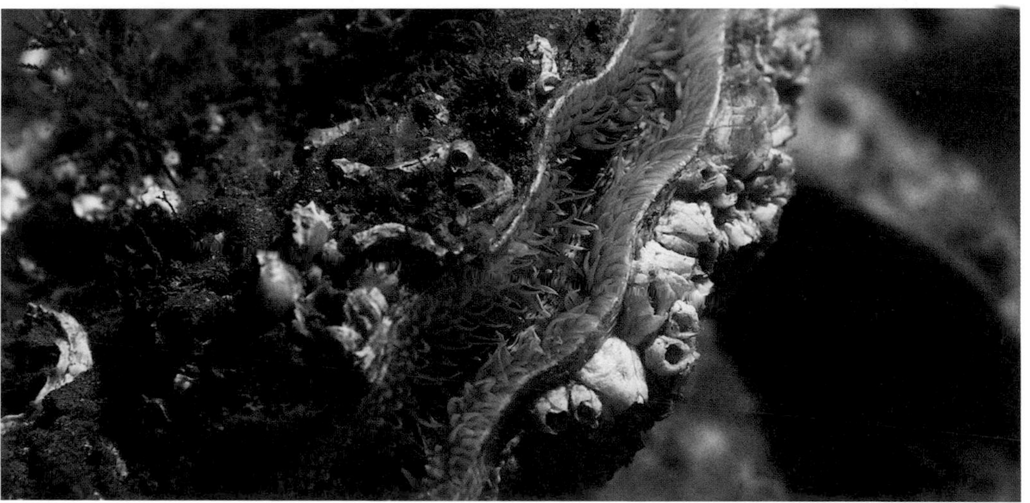

Hunters of the Deep

Once the horse mussel colony establishes itself on the seafloor, other life follows. Some are drawn in by the opportunity to hide and live among the mussels, while for others the hard shell of a mussel is perfect to attach to, and this abundance attracts those towards the top of the food chain.

The *Celtic Deep* dive team filmed octopus, sharks, fish and nudibranchs all hunting on the mussel beds.

The curled octopus, also known as the lesser octopus, is probably the most intelligent of all invertebrates found in Welsh waters. Elusive by nature, it spends most of the day hiding in crevices with its tentacles curled up – hence the name.

Like most octopuses it can release an inky dark fluid from its body when it feels threatened. The fluid creates a dark cloud in the water which confuses and disorientates predators, allowing the octopus to escape from danger. The octopus is a cephalopod (meaning 'head-footed'), from the same group of molluscs as squid and cuttlefish.

Growing to around 50 cm long, it is identifiable by the single row of suckers on its arms, in contrast to the common octopus, which has two. The curled octopus is an active predator, feeding mainly on crustaceans, molluscs and other invertebrates. When feeding on crabs, it immobilises its prey by

Below left: A red gurnard 'walking' the horse mussel bed on the hunt for small fish and crustaceans.

Below right: A curled octopus displaying its curled tentacles as it too 'walks' the reef.

Red gurnard.

Curled octopus.

Starry smooth-hound.

puncturing its eye or boring through the shell and injecting toxins into the body of the crab which paralyses it. The digestive enzymes contained in the saliva of the octopus break down the muscle within the crab's body, allowing the shell to be easily removed.

Stalking the horse mussel bed alongside the octopus is a red gurnard on the hunt. A curious-looking creature, the lowest three spines of the pectoral fin are separated and used to probe the seabed for hidden prey – they look a bit like legs and can give the impression that the red gurnard is 'walking'! These fish literally stalk the mussel bed looking for crustaceans and small fish.

Starry smooth-hounds also patrol the seafloor. Rather than the typical sharp shark-like teeth, the smooth-hound has blunt but powerful crushing plates which are adapted perfectly to consume the crustaceans that make up the majority of its diet.

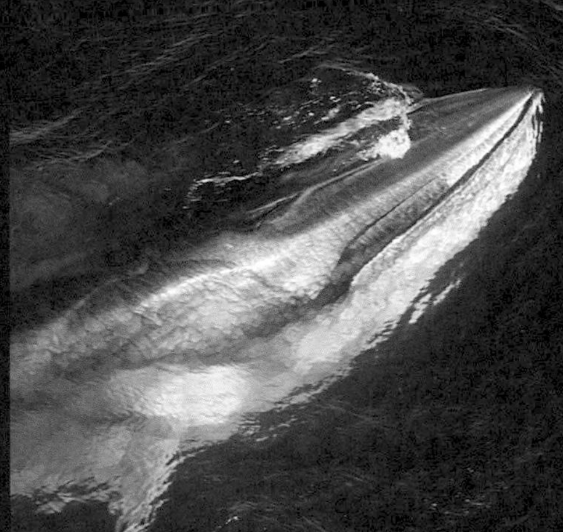

A fin whale with a mouthful of herring, filtering water and air out before swallowing.

Fin Whales

The fin whale is the biggest animal to be found in the Celtic Deep. At up to 25 m long, it's a true giant and the second largest animal on Earth. The only animal that can rival it for size is the blue whale, but they are so closely related that they have been known to successfully interbreed.

Before commercial whaling started in the North Atlantic the fin whale would have been a regular fixture in the Celtic and Irish Seas. Around a million fin whales were killed for their oil, meat and baleen in the early part of the 20th century, resulting in their near extinction. Today just 10 per cent of the original population remains, keeping them firmly on the endangered species list.

The number of fin whales visiting Welsh waters, along with their exact movements and migrations, is poorly understood. However, populations are on the rise and marine biologists believe that fin whales are now returning here on an annual basis.

Despite being found around the world and being hunted so heavily, the fin whale is little studied, seemingly hiding in the shadow of the blue whale when it comes to research.

Fin whales can be found throughout the world's oceans but are most at home in the deep, open ocean, preferring cooler waters in places like the North Atlantic, and are rarely found in tropical waters. Like other large whales, fin whales are thought to migrate between feeding and breeding grounds; however, resident populations are known to exist in both the Mediterranean and the Gulf of Mexico.

Fin whales are filter feeding, baleen whales like humpback, minke and right whales. Possessing 60-100 expandable pleats in their throat allows them to take in colossal amounts of water which they then sieve out using over 400 keratinised plates. A fin whale can consume up to 2000 kg of krill in a single day, but they will also feed on a variety of other small marine life such as other crustaceans, small schooling fish and squid.

When hunting schooling fish they lunge feed on their side at the surface – a behaviour filmed by the *Celtic Deep* crew for the final sequence of the programme.

Fin whales possess asymmetric pigmentation on their jaws: their right side is a brilliant white whilst the lower side of their jaw is grey-black. No one knows quite why this is, but one suggestion is that they use it as a counter shade when they roll onto their side during feeding lunges. Another idea is that they use the white on one side as a startle device to herd the prey into a tighter formation, and it may even be a combination of both.

Common dolphins bow riding off the nose of a travelling fin whale.

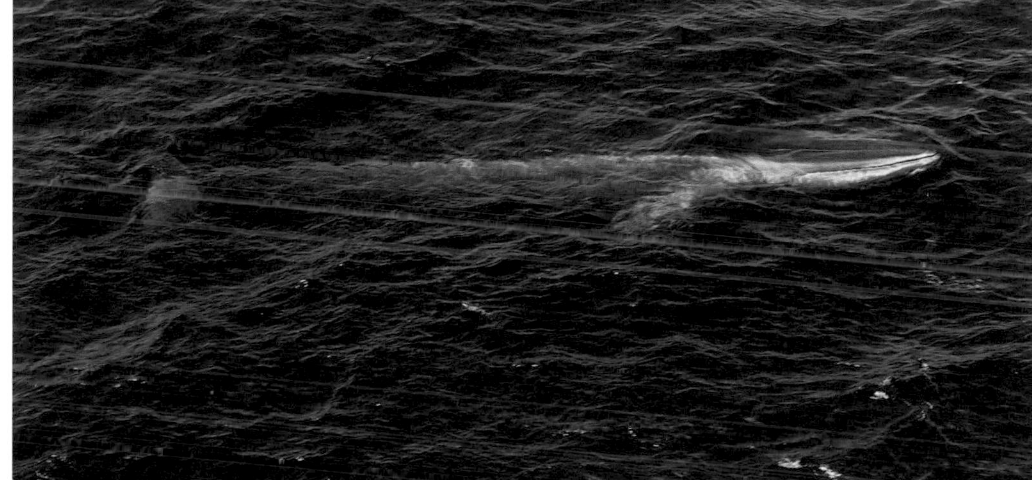

FACT FILE
FIN WHALE

- Latin name: *Balaenoptera physalus.*
- Can live over 110 years.
- Consumes up to 2,000 kg of krill a day.
- Can sustain a swimming speed of 40 kmph.
- Can dive to 450 m.
- Also known as finner; razorback; finback; flathead; common rorqual; herring whale.

A Feeding Frenzy

Pleats in the fin whale's throat allow it to expand when feeding.

By late summer the Celtic Deep is at its warmest and is full of life. The arrival of big schools of herring and other bait fish in the Celtic Sea triggers a feeding frenzy among both the residents and the seasonal visitors.

No one species has monopoly over the fish – especially if the birds have anything to do about it. Tuna are quick to arrive and demolish a shoal, but their splashing on the surface draws the attention of birds from miles around and the sight of hundreds of diving gannets makes a feeding frenzy easier to find than most other forms of marine wildlife behaviour, useful for any sightseers or film crews nearby.

Dolphins employ a slightly different tactic to the tuna's bombastic approach, swimming in circles around and under the shoal of bait fish, tail-slapping and keeping the fish in a tight ball against the surface until it's time to attack. As they pick off the fish the commotion draws in the tuna and others in the vicinity.

Fin whales, known for their speed, appear as if out of nowhere. They feed on their side, taking out much of the shoal in one massive mouthful. From a distance all that's noticeable of the whale is its enormous tail fin, travelling on its side as if mimicking a dorsal fin, but from above their sheer size dwarfs the shoal and the rest of the hunters.

The presence of the fin whales doesn't put off the dolphins and tuna, who merely dart around the huge animals picking whatever scraps they can. Even the smaller, but still large at seven to eight metres, minke whale almost slips in, taking big gulps as it swims by, just under the surface.

The bait fishes' defensive strategy of staying together in a tight ball is no use under this level of onslaught and the massive gape of the fin. Once this shoal is finished the diners will move on to the next one, leaving the few remaining herring to regroup.

This behaviour in its various forms and selection of species makes summer and autumn in the Celtic Deep busy and spectacular.

Above: Dolphins corall the fish close to the surface.

Into the Deep: Diaries

Filming in the Deep

For the *Celtic Deep* team this was to be the most challenging filming in the entire series. The difficulties presented in trying to film the intimate behaviour of marine wildlife in the open ocean meant that there was a very real possibility that, the team might not manage to film anything at all for this episode.

CRESSI
CRESSI
nauticam
SCUBAPRO

Filming in the Deep

The brief was to film Wales' pelagic open ocean species, many of whom are only here for a few months of the year, during one season of 2020.

Out in the Celtic Deep, the difference between being able to film or not is built on the slimmest of margins; tiny changes to the conditions could ruin the chances of filming anything, while a good forecast often did not lead to conditions good enough to film in.

The crew were at sea for days at a time, but often unable to film anything at all as it was either too windy or too rough to get the crew off the boat and into the water – it was even impossible to launch the drone due to a rocking deck. On those days the team would have to just watch the wildlife and their stomachs and head for shelter.

One beautiful hot sunny day the crew sailed out of Haverfordwest, hopeful that at least they'd be able to get in the water, only to spend the rest of the day enveloped in a thick sea fog on a slick, calm sea. They could hear whales breathing around them and regular frenzied splashes from animals feeding on bait balls, but could see nothing.

As evening approached and the boat needed to head for home, the fog started to lift and as the boat set off two curious fin whales stopped by as if to say hello before disappearing. They had been in the right spot all day, but with nothing to show for it.

Above: Series director Anne Gallagher preparing to drop the camera team off ahead of an evolving bait ball.

Another day started with promising conditions before a short period (meaning the waves were very close together) northerly swell appeared. Created by wind blowing through the Irish Sea it meant they were in for a very rough and long ride back. Again serenaded by a group of fin whales, the team could do nothing but watch and hold on tight.

On the few days when the conditions were perfect, the small challenge then arose of finding the wildlife in miles upon miles of open sea. It was hoped that a network of research boats and pleasure craft reports would provide up-to-date information on feeding hotspots where certain species might be found, but during the COVID-19 pandemic very few vessels were going to sea and data was almost non-existent.

With the help of knowledgeable local boat crews, tip-offs and a large dose of luck, the crew managed to witness and film a number of fin whales and the behaviour they needed to make the programme. As is often the way with these things, it came down to the last day and the last possible moment.

Above: Cameraman and Director of Photography Rob Taylor swims with only a snorkel to film common dolphins feeding on a bait ball of fish.

Heroes of the Celtic Deep

cressi

The Trouble in Our Seas

From leisure and tourism to the fishing industry, millions of people rely on the seas that surround Wales for both work and pleasure.

Left: 'Only a surfer knows the feeling', or so the old phrase goes, but it's only recently that that 'feeling' has been studied and shown to be scientifically beneficial to our mental health. Tonic has been working with Clynfyw Care Farm for a number of years and staff say they have seen great improvements in participants' mental health since joining in the surf sessions.

'As their confidence grows, and as their well-being improves, they start to come out of themselves. Seeing that benefit at the end, there's nothing better than that really.'

For Kwame, the mission is personal: 'It's been my saviour for my own mental health, if you like. I lost my father, my mother, my sister, all in a 12-year period, so, luckily for me, I had my own recovery tool.'

Tonic is now into its fourth year. Their mission statement is simple – who doesn't feel better when they go to the beach?

But what has this got to do with the conservation of our seas? Well, as Jacques Coustau, the famous pioneer of diving, said, 'people protect what they love.'

The Fisherman

For centuries Wales' coastal communities have lived off fishing and the sea, but gradually, as populations grew and fishing techniques evolved, so the decline of species followed.

Oysters used to be a Welsh speciality and oyster dredging became a key industry around Milford Haven and Swansea Bay during the 19th century. A large fleet was based out of Mumbles, peaking at 180 boats in the mid-1800s, but from there the oyster went into decline and by the mid-20th century its population and the trade it supported collapsed.

The herring industry has a similar story, with a thriving industry noted as far back as 1206. Nefyn in north Wales, Aberystwyth in mid, and Milford Haven in the south-west have all been important herring ports over the centuries, but a continual decline saw herring fishing all but disappear around the Welsh coast by 1970.

Milford Haven was one of the last herring fisheries open and saw more than 6,000 tonnes landed in 1925, making it the biggest landing station for herrings in England and Wales at the time. By 2005, just one tonne was landed there.

Across Wales today only a few hundred commercial fishermen are left, many of whom have had to take on land-based jobs to make ends meet.

Mark Gainfort from Dale in Pembrokeshire has witnessed the decline first hand: 'Back in the day there was plenty of mackerel. You used to be able to catch six at a time anywhere really, those used to be good days.'

Above: Mark Gainfort from Dale in Pembrokeshire endeavours only to fish what he needs after witnessing the decline in fish stocks over his lifetime.

Things have changed over the years and now the mackerel have had a bit of a hammering with the bigger factory boats.

'I used to do a lot of netting years ago. I tend to line catch as much as I can now. If you do catch the small ones on the rod you can put them back. [Now] I can go an 8 to 10-hour day and only get a few fish. You could have an 8 or 10-hour day and get no fish.'

Mark believes that consumers are becoming more and more conscious about where they buy their fish and hopes that we will see a shift towards more locally sourced produce, caught by more sustainable methods.

But for now Mark fishes slowly by line and pole, mainly for mackerel and bass but also pollock, skate, mullet, plaice and Dover sole, before selling his catch to the local pubs.

After the short walk up the pontoon, the fish is carried straight into the kitchen, where the chefs get to work filleting and preparing the fish dishes. It can be a matter of minutes before customers are served the catch. You can't get fresher than that!

Above: Fishing with a line and rod is one of the more sustainable methods of fishing, with little bycatch and allowing for the release of undersized and unwanted species. It is, however, commercially unreliable.

The Lobster Potter

The modern-day Welsh fishing industry is mostly based around shellfish, with 90 per cent of that being inshore – within six nautical miles of land. Of the nearly 2,000 fishermen around Wales, 1,000 of these are potters: fishermen who put out strings of baited pots on the seabed to catch crab and lobster. Much of this catch is destined for the continent, where Welsh crab and lobster are much prized.

The pots and traps used are highly selective for the species they target, with low rates of accidental catch. For the crab and lobster pots, incidental or accidental catch is primarily composed of undersized crabs and lobster and those that are soft or in poor condition, although fish can be caught too.

Despite this, lobster and crab potting is not without its impact on the environment. Lost pots can continue to catch crustaceans and fish for decades to come, with each caught animal becoming bait to bring in more – often called 'ghost fishing'.

Pots which aren't lost can catch multiple animals before being checked and younger and weaker animals provide food for others also trapped.

Brett Garner, a lobster potter based on the Llŷn Peninsula, has been fishing inshore for over 30 years and he's seen the lobster population dwindle over his career: 'I started with a 14 ft Mirror dinghy. I was catching 17 lobsters a day, out of 23 pots. I'd die for that now.'

Already undertaking the recommended practices for a sustainable lobster fishery, including the throwing back and v-notching (marking the tail) of berried hens (egg-laden females) and soft, freshly moulted lobster, Brett was looking to do more. So he linked up with Bangor University and was given access to their latest fisheries research. The university had been trialing putting escape hatches on lobster pots, square holes designed to allow fish and undersized lobster to free themselves from the pots.

Skeptical at first, Brett thought he had nothing to lose, so gave it a go and was soon won over: 'So I put 25 of these gaps in and I looked at them and I honestly expected this wasn't going to work, but the following season every pot has got them in, that's how effective it is.

'It's a no-brainer to me. It saves you time cleaning, you get done quicker, less time on the sea, less time working, less time measuring, because pretty much everything you've got [in the pot] is of [legal landable] size. Then there's been no fighting, so it prevents them from getting mutilated, losing their claws, which lowers the value at a later date when you catch them of size. And it stops things getting killed in the pot and also reduces unwanted fish.

'And the other thing that it most definitely does is it thins out ghost fishing. So if you lose a string, the [smaller] lobsters are able to get in and out. So it doesn't stop it. But it reduces it.'

Brett is hopeful that others can take up these practises to help secure the Welsh lobster population and the potting industry for the future, but he also thinks we all have a responsibility for and a part to play in the fish and shellfish industry: 'The consumer is the biggest element in it. We need the consumer to put the pressure on the fishermen to fish sustainably.'

Brett is continuing to work with Bangor University to learn more about the Llŷn lobsters: 'With Bangor [University] we tagged 300-400 and out of all the lobsters which were recaptured they only moved in a region of 300 m, so it really highlighted that if you look after your own patch, it should look after you.'

Often expected to be red in colour, the common lobster (*Homarus gammarus*) is in fact a bluey-black colour on top with a pale underside.

Seal Pup Rescue

If you were to pick an animal to represent the county of Pembrokeshire it would probably be the Atlantic grey seal. This charismatic and curious mammal is found in every nook and cranny of Pembrokeshire's coastline, especially in late summer and autumn when the pupping season starts.

The young pups are unable to swim when born and dependent on mum for the first three weeks. Between 10 per cent and 30 per cent of pups don't make it through these crucial weeks, with some years having much higher mortality than others.

Two common threats to young pups are disturbance and storms. Many of Pembrokeshire's pupping beaches are inaccessible, but for those seals who pup on beaches with access for people there's a real threat of the mum being disturbed to the point of abandoning her pup.

Even the most remote beach has its dangers. West Wales in the autumn is susceptible to Atlantic storms and the further into the autumn a pup is born, the more likely it is to encounter one. Large swells and big autumn tides leave the white-coated pups vulnerable to being washed off their normally safe beaches.

Luckily for the seal pups of west Wales there's a small team of volunteers ready and waiting no matter the weather.

For over 30 years Terry Leadbetter has been rescuing washed up, injured and abandoned seal pups. Now in his 70s, he's had a busy retirement: 'Well, my autumns are spent mainly listening to the phone and waiting for calls from members of the public and whoever to tell us that there's a seal on a beach somewhere. We don't have much time for anything else during the autumn.'

Operating as the charity Welsh Marine Life Rescue, Terry and a small team of volunteers, including his neighbour John Beavis and wife Ann, who deals with the medication and husbandry of the seals, rescue around 35 pups a year but are called out to many more.

Above: Terry Leadbetter (left) and John Beavis (right) with their trusty quad bike, allowing them to rescue seal pups from remote beaches and coves across Pembrokeshire, Ceredigion and Carmarthenshire.

'Often we go out and there's nothing wrong with them at all and they don't need to be rescued, but if there is a problem with them or they're orphaned, and we can clearly see that they're orphaned, we'll pick them up.'

In a bad year, such as 2017 when Hurricane Ophelia and Storms Aileen, Brian, Caroline and Dylan all passed through during the pupping season, this can result in a big loss of seal pups; though a lucky 76 pups passed through the rescue team's hands that year.

Once Terry and John pick up a seal in need of help, they bring it to a small facility at their home near Milford Haven, while Ann prepares food and any medication needed. The youngest pups will still be being fed by their mother and, unable to replicate a mothers' milk, Ann makes a fish smoothie which is then fed to the pup via a tube.

They are able to keep seals at their home for more than a week, conducting what Terry calls the triage phase while waiting for a space to be available at one of the RSPCA's rescue centres. The RSPCA then takes care of the long-term rehabilitation and release. It's a good partnership, he says, 'Unfortunately, the RSPCA have got such a variety of animals to deal with and they can't be in the same place at the same time. So the fact that I can pick up seals and keep them until they've got the availability to take them, working together, it works really well.

'If anybody does come across a seal on a beach, don't try to be a hero; keep your distance from it and keep kids and dogs away from it. Don't cause it stress. If you overcrowd it, and it's a little white pup, there is a possibility that the mother would be able to come and feed it, but she won't because there's too many people about. So the best thing to do is just call us anyway and we'll come out and check it.'

Bottom left: During the autumn, Welsh Marine Life Rescue receives hundreds of calls and can have multiple trips out for the same animal. If a seal pup is in healthy condition with a mother nearby, it is marked to prevent further call-outs.

Stranded

Every year, between 350 and 800 cetaceans (a collective including whales, dolphins and porpoises) wash up on beaches around the UK. Most are dead, but some are still alive.

On 12 June 2020, a young fin whale was spotted alive on a sandbar in the Dee Estuary. One of the first on the scene was local volunteer Gem Simmons of British Divers Marine Life Rescue aided by the RNLI, local fishermen and other volunteers.

British Divers Marine LIfe Rescue, (BDMLR), whose mission is to provide assistance to any aquatic (marine and freshwater) animal in need of help, train volunteers to be able to respond to stranding events in their local area.

Sadly, live whale strandings rarely end with a successful return to the sea, especially when an animal as large as a fin whale ends up stranded – their large bodies are not designed to support their weight on land.

In the case of the Dee fin whale there was little the team could do but make it comfortable with water and wet towels to keep its skin damp.

The Natural History Museum is responsible for monitoring cetacean strandings under the UK Whale & Dolphin Stranding Scheme. Since the scheme started in 1913, more than 11,000 animals have been recorded and it is one of the longest recording schemes of its kind, providing a wealth of data.

Gem Simmons (right) timing the young fin whale's breathing. This helps to establish its condition and stress levels.

Museum scientists study the remains of dead stranded cetaceans to learn more about their biology. The scheme records how many cetaceans strand in Britain each year, what species they are, where and when they strand and the age and sex of the animals. They also research animal behavior and uncover causes of death. Researchers are then able to use strandings data to track whale population numbers.

Paradoxically, if more dead whales wash up on beaches, it is often a sign of a thriving population. If no cetaceans are spotted (even dead ones), it may indicate their numbers are struggling.

In recent years the numbers of cetacean strandings being recorded have been on the rise. This might not sound like a good thing, but could be down to a number of positive factors.

The first is simply that there are more people aware of the UK's marine life and recording what they find on the beaches. The ubiquitous use of smartphones means that not only are people more likely to positively identify what they find, they are also more likely to report it to the correct organisations.

It could also represent an increase in the overall numbers of cetaceans now living around the British Isles, although this is only speculation and not easy to prove, but there has been a corresponding increase in cetacean sightings – so maybe there is a good news story behind it all.

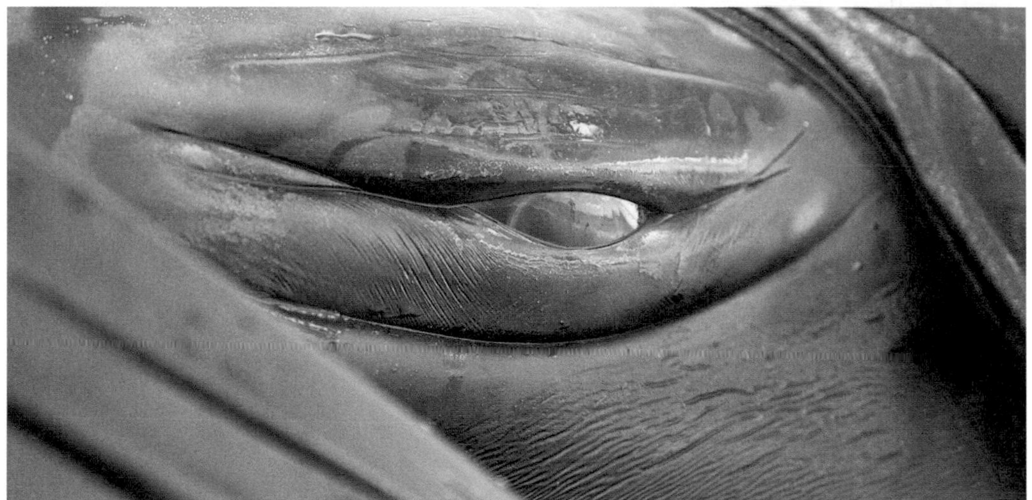

Right: The skin on a whale is delicate and needs to be kept moist at all times. The BDMLR volunteers cover the whale in wet towels and form a production line to continuously bring buckets of water to keep the whale comfortable.

Citizen Scientists

Despite modern technological advances, such as the use of drones and satellite technology helping scientists have unprecedented views of remote environments and their wildlife, much of our marine environment around the world remains little visited and rarely studied.

The waters around the Welsh coast are no different, with big gaps in knowledge about the species that live here and their behaviour. But there are those that are helping to try and fill in some of these gaps, and they aren't all PhD-holding scientists. Instead they are teams of ordinary people – citizen scientists – all playing a small part in monitoring the environment they know and love.

Citizen science is defined as the practice of public participation and collaboration in scientific research to increase scientific knowledge. Through citizen science, people share and contribute to data monitoring and collection programmes. Usually this participation is done as an unpaid volunteer.

Citizen science projects are as varied as the people that contribute to them, and may include wildlife-monitoring programs, online databases, visualization and sharing technologies, or other community efforts.

For one Pembrokeshire teenager a fascination with rockpooling and exploring the intertidal zone of his local beaches has led to an in-depth citizen science project.

Above: Aran (left), Sam, Hamish, Charlie, and Brinley are teenage citizen scientists studying and logging the wildlife found in the intertidal zone of Dale beach, Pembrokeshire.

But a project like this is best done as a team, so Aaron Lock is now joined by his friends Charlie, Sam, Brinley and Hamish for the seasonal surveys.

In each survey, conducted around the biggest low tides of the season, they try to find as many species on the shoreline as possible. They then repeat this at different times of the year to see if the species present varies by season.

Their methods aren't complicated, mostly searching by hand, turning boulders or sweep netting with a large homemade net to catch the wildlife that's too quick to spot. Everything that is found is logged and submitted to a national database which can be accessed by those doing further research into the diversity of wildlife and particular species found on our shorelines.

Aaron really believes in the importance of their work: 'It can't just be reliant on the marine biologists to constantly come forth with records. So if someone takes it upon themselves, as a hobby, to get a recording of things they can find at one of their local beaches, that can easily go into the databases that can be used by scientific teams who want to understand a little bit more about our coastlines.'

In the UK there are plenty of easily accessible citizen science projects for those with an interest in wildlife. Popular projects include the RSPB's Big Garden Birdwatch and Shoresearch and The Wildlife Trusts' national citizen science survey of the intertidal shore. If there's something you're interested in then there's probably a research project you could contribute to as a citizen scientist.

Above: Using a large sweep net in the shallow water allows for the safe catching of creatures such as scorpion fish and conger eels.

Jake's Mission

Wales' underwater world is unknown and inaccessible for much of the population, but one marine biologist is determined to change that.

Scuba diver and sailor Jake Davis grew up on the Llŷn Peninsula, often getting up at 5am before school to help his dad with the lobster pots. These days he's more likely to be in the water himself or conducting research by dropping a BRUV – baited remote underwater video camera – into the sea.

After a childhood of fishing and snorkeling evolved into scuba diving and a marine biology degree, Jake worked for the Welsh branch of the international Angel Shark Project to discover if this rare shark is still to be found in Welsh waters and if it might be home to their nursery ground.

An ambush predator normally found submerged in sandy habitats, lying in wait for unsuspecting prey while camouflaged against the seafloor, the critically endangered angel shark was once common along the Welsh coast.

Despite suffering widespread decline over the last century, there have been an increasing number of sightings in recent years, giving hope for its future.

For Jake, after the research, one of the most important aspects of his work is to share the wildlife he sees daily with the general public and inspire in them the passion he has for our underwater world.

Bottom left: A BRUV – baited remote underwater videocamera – allows Jake to gather footage of shyer species which aren't normally seen by divers.

As well as looking for the angel shark, his underwater cameras document daily life in our seas without the presence of divers or other disturbance. Once the footage has been logged for the research, Jake picks the best moments to share with his followers across social media. He especially like to share the footage of sharks:

'A lot of people don't really think we have sharks off the Welsh coast and all of them are completely harmless species. So last year, I put one [clip] up of two tope sharks swimming around. And it went almost global because [people were saying], 'This can't be the Welsh coast', but it is and it's really nice to show people that we do have these species. It's positive engagement for Welsh marine life, UK marine life, and giving sharks a positive press for a change.

'This [footage] gives us more information and is really good for scientific uses, but it's also really powerful when shared on social media. People then get to see it for themselves. It gives them that appreciation of what they have on their own doorstep.'

For Jake, this sharing of footage and education of the wider public is as important as the research itself – for people to love it, they must first know it.

A common lobster (middle) tries to work out how to reach the bait.

Frankie Hobro in the seahorse breeding facility at Anglesey Sea Zoo.

Saving the Seahorse in Wales

It may come as a surprise to the beach-going public that Wales has two native species of seahorse, *Hippocampus hippocampus* (short-snouted seahorse) and *Hippocampus guttulatus* (long-snouted seahorse).

Both species used to be found in shallow waters all around the Welsh coastline, perfectly camouflaged in seagrass beds and rocky reefs. However, the last confirmed sighting of a seahorse in Wales was off Newquay around 20 years ago, so they are now believed to be extinct in the wild here.

Seahorses are under threat worldwide and the demand for seahorses and seahorse-related products is vast, with over 77 nations trading 25 million seahorses each year. Because of this, seahorses are now on the IUCN list of threatened species. Much of this demand is for the home aquarium trade, but as seahorses are very fussy about what they eat and need a constant supply of live food and specific living conditions, most aquariums are unsuitable for them, so sadly they do not survive for long in captivity.

Despite being such popular animals, very little is known about their numbers and behaviour. There are over 47 species of seahorse across the world, with new species being discovered every year. Five new species of pygmy seahorse were discovered in 2009 in the coral reefs of the Red Sea and Indonesia.

In the UK in 2008 the Countryside Act was passed to protect British seahorses and it was acknowledged that both species were data deficient. Very little was known about their numbers and distribution around the British Isles, except that they were once anecdotally seen around much of the shallow coastal areas but then almost entirely disappeared over a decade or two. Having so little information made it extremely difficult to know why they had suddenly gone into decline, so it was vital to gather as much information as possible in order to protect them.

At this crucial juncture, Frankie Hobro, owner and director of Anglesey Sea Zoo, stepped up to the challenge of not only rescuing the two species but learning as much as possible about this little-known animal in a natural captive environment. In order to gather accurate information it was essential to create captive conditions and represent those in the wild as closely as possible, but with seahorses being so

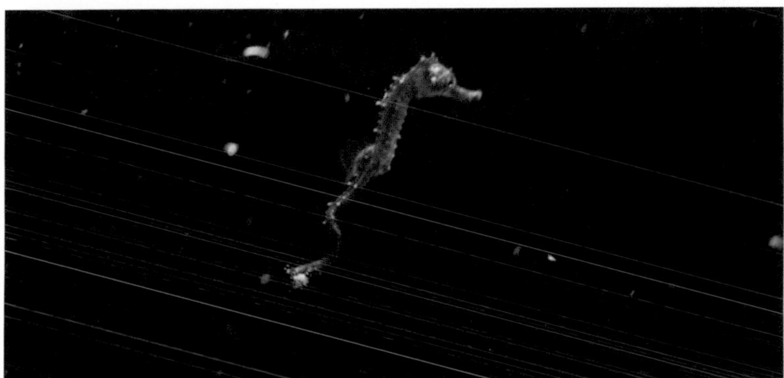

elusive and tricky to keep, this was a huge challenge!

'We homed some of the last wild-caught short-snouted seahorses, *Hippocampus hippocampus*, here in 2008 and started to try and breed them and there was no information on what to do, so nobody could really give us any advice, other than what hadn't worked.

'It took us three years of dedicated husbandry techniques, close observation and trial and error to successfully raise some juveniles and, with the techniques that we'd used, we were then able to replicate those with ZSL (Zoological Society of London). The following year they managed to raise some seahorses as well from the same wild-caught batch and those were distributed through different aquariums.

'So while we were successfully breeding them up, they were still captive stock at that point, there weren't any more wild-caught animals [being found].

'In 2015, we were able to get funding to get (and breed) the second species of British seahorse, *Hippocampus guttulatus*, and again there was very little information on how to keep or breed this species.

'We are working closely with other organisations, including Project Seagrass, Seasearch and local fishing communities, to gather as much information as possible on the habitat requirements of both species and their previous distribution to ensure that we have suitable release areas ready for when we are able to start re-releasing.

'So everything that we do is a really important learning experience not just for us in husbandry with the animals but also with a view to re-releasing them.'

Now Frankie and the team at Anglesey Sea Zoo are successfully breeding both native species of seahorses, their next challenge is to acquire some wild-caught animals to create a more robust gene pool, so that the next generation of young seahorses can be released into the wild with the highest possible chance of survival.

These two species of Welsh seahorses are not out of the woods yet, but Frankie hopes that within the next decade they will be breeding seahorses for release and once again there will be healthy populations of seahorses re-established in the wild around the coast of Wales.

The Paddleboarding Plastics Campaigner

Marine litter pollution is a global issue and Wales has its part to play in both contributing to the issue and trying to solve the problem. Over 80 per cent of the litter that enters the marine environment each year comes from land-based sources, making its way to the sea through our drains, rivers, canals and other waterways along with materials flushed down toilets and waste from manufacturing products.

Sian Sykes from Anglesey is a stand up paddleboarder and campaigner against single use plastics. With plastics being so common in our lives, around 70 per cent of all litter in the oceans is made of plastic, which doesn't break down.

To highlight the plastics issue, Sian undertook a record-breaking circumnavigation of Wales on her paddleboard, a 1,000 km journey over 60 days. Her campaign helped make Anglesey the UK's first 'plastic-free' county, with a significant drop in single-use plastics.

She's driven by the amount of plastic pollution she sees out on the water: 'We're so fortunate to have such a beautiful coastline. It's resilient. The tide comes in twice a day, and it feels like a new beginning. But we're taking advantage of it. We're polluting it. We're not being kind to it. When I see plastic bottles floating on the ocean, it just breaks my heart that it's us humans having an impact on this beautiful, beautiful environment.'

Left: Record-breaking paddleboarder Sian Sykes campaigns against single-use plastics.

'The general public thinks plastic is litter that somebody just dropped, litter in the street, but it's a lot more than that. When the plastics end up in the ocean, they're breaking down into lots of micro plastics. And that's getting into our food chain.'

Microplastics are tiny particles of plastic that have disintegrated from larger pieces. These microscopic plastic particles might be difficult to see but they arguably have an even bigger impact than large pieces of visible plastic. Toxic pollutants found in the environment have been shown to be attracted to the surface of plastic pollution and have the potential to impact human health through the food chain.

As Sian explains, 'It's estimated 51 trillion pieces of microplastics are floating around. A lot of fish think it's food, and then we're eating the fish. I was heartbroken to find that there's been studies showing that microplastics have been found in [a human] placenta.'

The wildlife is suffering too – postmortems on stranded whales and turtles often find stomachs full of indigestible plastics. The same has been found in seabirds, a recent study finding that 93 per cent of the fulmars in the North Sea have some plastic in their stomachs, while one bird was found to have ingested 454 pieces. Wildlife of all sizes is starving to death with stomachs full of plastic.

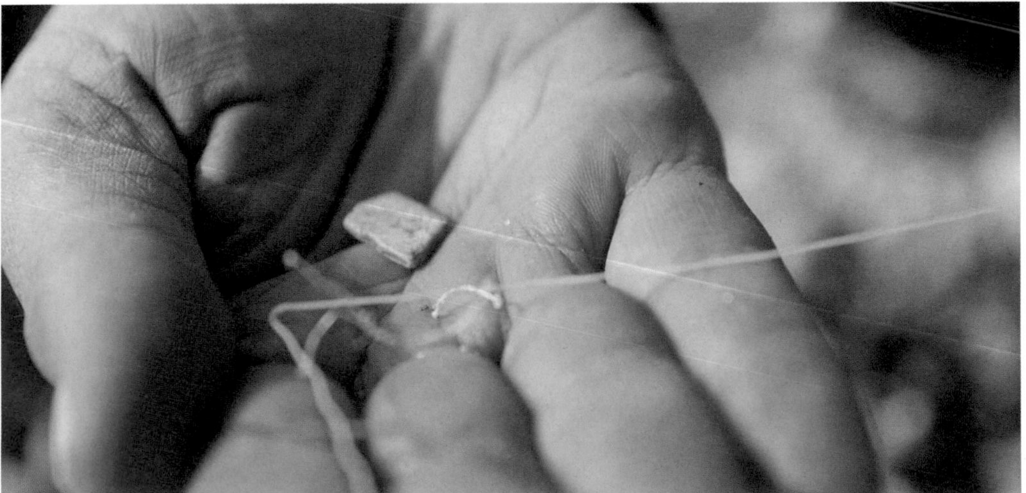

Right: Around 70 per cent of all litter in the oceans is made of plastic. When plastics 'break down' they just become smaller pieces, known as 'microplastics', which can end up being eaten by fish and entering the food chain.

Underwater Litter Pickers

The litter and debris we see washed up along our beaches and coastline are just the tip of the iceberg when it comes to marine litter. Much of the impact of our marine pollution happens in remote and difficult to reach areas of our marine environment.

In 2005, Pembrokeshire plumber and scuba diver David Kennard was having a few beers with friends after taking part in a beach litter pick when he had the idea to set up the UK's first underwater clean-up group. Neptune's Army of Rubbish Cleaners, or NARC, was born.

Over 2,000 clean-up dives later this award-winning charity also runs anti-pollution campaigns and educational days but the underwater water clean ups remain at its core.

The group of over 30 volunteers from every walk of life tackle all sorts of marine litter but keep coming back to dive sites notorious for two types of pollution – fly tipping and ghost fishing. Corners of the coastline with easy access to the water are unfortunately easy targets for fly tipping. The divers have retrieved all sorts of discarded items from cars to toilets and DVD players.

But by far the biggest aim of the NARC team is to reduce the amount of abandoned and lost fishing gear, also known as ghost fishing gear, in Pembrokeshire's waters.

For our marine wildlife lost fishing gear is a deadly form of litter, as it indiscriminately continues to do its job of catching wildlife, entangling fish, marine mammals, seabirds, and sharks.

Above: David Kennard (front right) heads up NARC, a team of over 30 volunteers all willing to take on the cold and difficult task of removing human debris from the seas off Pembrokeshire.

Right: Lost nets, lobster pots and fishing line end up ghost fishing and trapping wildlife for many years to come – unless the NARC divers turn up to remove it.

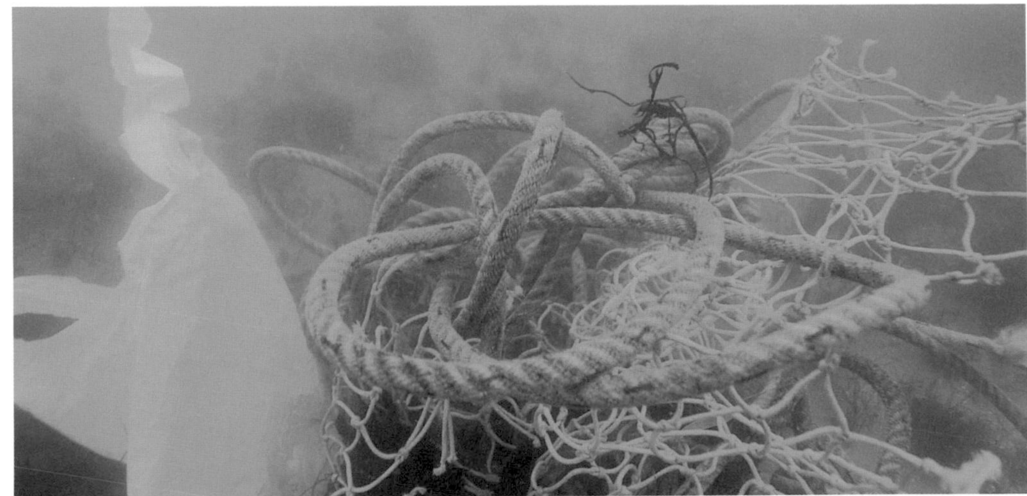

Often the trapped animal inadvertently becomes bait for the next.

Ghost fishing gear also damages delicate marine habitats, catching on delicate corals, sponges and seaweeds which can take decades to grow. Additionally, it's responsible for the loss of commercially valuable fish stocks, impacting on both the overall sustainability of fisheries as well as the people who depend on fish for food and livelihoods.

As the work of NARC became more known across the county, fishermen and potters now contact the group with GPS points to alert the divers to areas where gear has been lost. A partnership that works well for both fishermen and wildlife.

It's not just commercial pots and nets that cause ghost fishing problems. Sea angling is one of the most popular pastimes in the UK but broken line and hooks can continue to catch wildlife for hundreds of years to come, making retrieving it a priority for the NARC team. Lobsters, fish and sharks are all freed by the divers – with minimal thanks from those released.

A Changing Climate

The scientific consensus is that climate change is happening and that climate-warming trends over the past century are extremely likely to be due to human activities. Increasing emissions of carbon dioxide and other greenhouse gases, as well as land use change, are the processes primarily responsible for the increase in warming.

Climate change is happening, but for many its effects still feel no more than a vague threat on the horizon and manifest as a particularly hot summer or an even more stormy winter and news stories from far-off places. For others suffering from back-to-back flooding events in the same year or losing crops and livelihoods to unseasonable weather the impacts are all too real.

Scientists estimate that sea levels could rise by more than a metre within the next 80 years, driven by melting glaciers and the thermal expansion of sea water as it warms.

This rise will put many of Wales' coastal communities at risk, and some are already struggling after winters of increasingly violent storms battering their sea defences.

Governments, policy makers and local councils are now starting to make decisions on how we face these threats and changes as they approach, and some are already starting to feel the effects of these decisions.

Above: The village of Fairbourne in Gwynedd is poised to be one of the first Welsh communities to fall victim to rising sea levels. In 2013 the local council announced that it could not afford to maintain their sea defences indefinitely.

One of those communities at risk from sea level rise is Fairbourne in Gwynedd. Built on saltmarsh on the edge of the Mawddach estuary, it's sandwiched between the mountains and the sea and kept dry by sea defences, tidal defences and a network of ditches.

In 2013 Gwynedd council decided that it could not afford to defend the village indefinitely and announced that maintenance of the sea defences would stop in 2054, concerned that defending it in the face of further sea level rise would be a losing battle.

Angela Thomas, Clerk of Fairbourne Community Council, moved here in 2013 to a house just behind the seawall.

'It's very easy to see why people are attracted to this place, what could be more beautiful? We have Cardigan Bay behind us. If you look further afield you can see Cadair Idris, we have the mountains, we have the estuary.

'When we know that we've got a high tide coming in, and that there's going to be a wind behind it, there's no doubt that we are worried. We can lie in bed at night and hear the swish of the sea on the shingle and wonder how high this tide is going to be.

'It is very hard to imagine that the sea could rise so dramatically in years to come, that Fairbourne will have to be decommissioned. But realistically, sometime in the future, we could possibly see the demise of Fairbourne and many other communities on the north Wales coast and the rest of the UK coast.'

Fairbourne might be the first community to have this decision foisted on them, but it's unlikely to be the last. The government-funded UK Climate Change Impacts Programme suggests that by the end of the century sea levels around Wales will have increased by up to a metre.

With over 60 per cent of the Welsh population living in coastal towns and cities, this puts most of the population within reach of the impacts of rising sea levels and the coastal erosion that comes with it.

Angela Thomas (left), Clerk of Fairbourne Community Council, moved to Fairbourne in 2013 to a house just behind the seawall.

Dr Richard Unsworth (front centre) and some of the many volunteers and scientists that help make Project Seagrass happen.

Project Seagrass: Planting the Ocean's Rainforest

The sheltered bays and estuaries of Wales may play a surprisingly important role in the global fight against climate change. Seagrass beds – a much overlooked habitat – could help in both fighting climate change and restoring the health of our seas.

Seagrasses are the only flowering plants able to live in seawater and pollinate while submerged. They often grow in large groups, giving the appearance of an underwater meadow. Like most plants, they depend on high levels of light for photosynthesis to grow and can therefore only be found in shallow water to a depth of around four metres.

Sadly, since 1980 Wales has lost 35 per cent of its seagrass meadows and across the UK the picture is much worse, with a loss of 92 per cent. These statistics become even more worrying when set next to the advantages this simple habitat provides.

Seagrass absorbs CO^2 35 times faster than pristine rainforests and it has been calculated that seagrass across the world is responsible for 15 per cent of the ocean's total carbon absorption. In return, as with all plants, it releases oxygen, with one hectare producing approximately 100,000 litres of oxygen per day.

Against rising sea levels and increased stormy conditions, seagrass beds slow down wave action and help stabilise the seabed with their roots, slowing erosion. Seagrass meadows are also a haven for wildlife, acting as a nursery for schools of young fish and, as a result, support many of the world's biggest fisheries. Just compared to bare sand, a seagrass bed increases animal biodiversity by 30 times. One hectare of seagrass could provide a home for approximately 80,000 fish and 100 million invertebrates – just think how many seahorses could find a home in a healthy Welsh seagrass meadow.

Losing seagrass, therefore, would be a huge loss to the planet, and could exacerbate damage done already, but still this delicate habitat continues to suffer from destruction, pollution and development. In some tourist destinations it is thought to be unsightly and pulled up.

Luckily, this habitat underdog has a Welsh champion in the form of *Project Seagrass*.

Born out of Swansea University and set up by university lecturers Drs. Benjamin Jones, Richard Unsworth, Richard Lilley and Leanne Cullen Unsworth, the project has made the replanting of beds of seagrasses in the Welsh waters their top priority.

Project Seagrass was created to take cutting-edge research out of the lab and into the field and make a difference to a habitat they had fallen in love with and could see was at risk of disappearing. The modest aim – to plant a million seeds and restore one of Wales' seagrass beds.

The first phase was to hand harvest the seeds from a healthy seagrass bed using snorkel or scuba, a very slow process reliant on many volunteers giving up often weeks of their time.

These stems are then taken back to the lab for the outer stem to rot away and the seed to harden as it would in the wild. The seeds are then placed by hand into small sand-weighted hessian sacks before being tied together and planted in chains along the seafloor of an appropriate bay, one that might have held seagrass in the past or whose seagrass bed is much depleted.

Their initial aim was to plant a million seeds in sheltered Pembrokeshire bays with the aim of restoring acres of this underwater meadow, and with that target now within reach and fresh seagrass shoots appearing in planted locations, it's going to be on to the next challenge.

Once Pembrokshire's seagrass beds are re-established they hope to branch out across the rest of Wales and the UK before trying it in damaged seagrass meadows across the world, saving our seas and ultimately our planet one seed at a time.

Right: From collecting the seagrass seeds by snorkel to individually bagging the seeds up to plant back out in a sheltered bay, the work of *Project Seagrass* takes many dedicated hands.

The Future

The threats to the seas around Wales continue to be many, varied and, for the most part, human driven. But we are a resourceful species and, as these pages show, are full of drive and ingenuity for solving the problems we face.

For each of the heroes featured in *Wonders of the Celtic Deep* there are many more not written about. Around the Welsh coastline there are thousands of people quietly doing their bit for their patch, from the dolphin spotters and loggers to the beach cleaners and no plastic users. No one can solve it all, but with many people following their passion and using their expertise, maybe we can hope to turn the tide and restore health and biodiversity to our seas, and the benefits to our own health and well-being surely make this an easy choice to make. The future is in our hands, our collective hands – together, we can make a difference.

If someone here has inspired you to action but you aren't sure where to start, look around your local community for groups and individuals already running projects you can join. Alternatively, there are large groups and charities working on a local level to help our marine wildlife.

If there's nothing suitable for you, why not think about your own daily life and what small changes you might be able to make to help improve the health of our oceans. There are lots of online resources on how to make simple, sustainable changes to cut down on your carbon emissions and reduce the amount of waste you produce or plastic that you use.

For the more landlocked, every waterway leads to the sea, meaning your actions in your local area are every bit as important as those done on the coast.

Here are just a selection of groups working around Wales and the UK, some of whom were featured in this chapter:

- The Wildlife Trust
- Marine Conservation Society
- Sea Trust Wales
- Seawatch Foundation
- Seasearch
- Surfers Against Sewage
- Beach Academy Wales
- Welsh Marine Life Rescue
- British Divers Marine Life Rescue
- Anglesey Sea Zoo
- Project Seagrass
- #2minutebeachclean
- Neptune's Army of Rubbish Cleaners (NARC)

All these groups have an online presence and there are many more not listed, including local, small-scale and more informal groups.

The Team

Dale Templar
Executive Producer and Managing Director of One Tribe TV

Anne Gallagher
Series Director

Sally Weale
Series Producer

Dale is the Managing Director of One Tribe TV and the creator and driving force behind the series, a project that has taken two years to make. She has worked as a TV producer and director for over 30 years; passionate about the natural world, she spent a decade at the BBC Natural History Unit in Bristol. Dale has filmed in over 60 countries, having travelled from the North Pole to Antarctica. She started working in Wales in 2011, while making the BBC multi-BAFTA award-winning landmark series *Human Planet.*

Wonders of the Celtic Deep was the dream job for Welsh director Anne Gallagher. Hailing from Porthcawl, she's a keen diver who grew up in the rockpools and surfing the waves of south Wales. After going to Aberystwyth University to study geography and film and go surfing, she started making her own wildlife films before eventually getting a job at the BBC's Natural History Unit. Her recent work includes the series *Wales: Land of the Wild.*

Sally series produced *Wonders of the Celtic Deep*. Prior to this she made documentaries for the BBC, National Geographic and the Discovery Channel. For several years she ran the Natural History Museum's Film Unit, leading a team making films for its exhibitions and digital channels. She first learned about the wonders of the Welsh coast listening to her mother's Dylan Thomas records. Passionate about wildlife, she studied natural sciences (at university) and is a trustee of the South and West Wales Wildlife Trust.

Dr Rohan Holt
Consultant and Safety Diver

Rob Taylor
Director of Photography

Jet Moore
Location Manager

Rohan graduated from the University of Liverpool with a degree in marine biology followed by a PhD based at the University of Glasgow, where he developed his expertise in benthic ecology and a love of underwater photography. This led to working for the Joint Nature Conservation Committee in the UK and other agencies abroad – finding areas of marine conservation importance and documenting their species and habitats. He continued to work for the government's conservation agencies for over 30 years to pioneer methods for monitoring marine habitats and species and eventually set up his own business to combine underwater wildlife filming and land and aerial photography with marine biological survey work.

Rob Taylor loves the water. As a child he spent holidays camping in Pembrokeshire and Morfa Nefyn and as a teenager would whitewater kayak in Snowdonia and at the Bitches in Pembrokeshire.
Since then he's travelled the world getting wet for broadcasters such as National Geographic, Discovery Channel and the BBC, but had rarely dived in the UK, so he jumped at the chance to spend a summer diving and working along the Welsh coast and loved every day in the water, from cold nights to pre-dawn summertime starts. He hopes he's done the Welsh coastline proud.

Jet Moore set up Adventure Beyond at the age of 18 to inspire people of all ages to enjoy and respect the natural environment. He has lived in Wales all his life and spent years learning about and exploring the coastline, but the opportunity to support the making of *Wonders of the Celtic Deep* was a once-in-a-lifetime experience. His skills in adventure sports and knowledge of the area, along with a keen interest in its flora, fauna, wildlife and geology, enabled him to lead the team safely through each shoot, bringing in additional support where required. He took huge pleasure in working with so many incredible and inspirational people, from marine biologists to contributors and everyone on the ground in the office, and is proud they managed to pull off what other productions would have three years to do in less than a year.

Acknowledgements

We would like to thank the following for their contribution to the *Wonders of the Celtic Deep* series:
BBC National Orchestra & Chorus of Wales, Creative Wales, Orange Smarty, Jessie Anderson, Nick Andrews, Daniel Attwood, Francesca Barbieri, Kara Bassett, Jack Carey, Julian Carey, Darran Clement, James Cox, Matthew Cox, Ilisa Factor, Mark Ferda, Owen Gay, Hannah Gosney, Tom Hanner, Christina Hill, Rob Hill, Graham Horder, Cameron Howells, Joe Hufford, Sam Hui, Summer-Anne Kiernan, Robin Lewis, Kate Lock, Daf Matheson, Benjamin McMillan, Paul Mealor, Nicki Meharg, Lindsay O'Brien, William Osman, Dame Siân Phillips, Luca Pittalis, Jessica Pollard, Sophie Pollock, Justine Rebello, James Reed, Martin Sampson, Jake Smallbone, Richard Stevenson, Lauren Thompson, Olivia Tonkin, Matt Waddleton, Samuel Webb, Steve White, Jesse Wilkinson, Aaron Williams and Dan Winkler.

Photography

Alamy: Front cover (top), endpapers, p. 3, 6, 12-13, 18, 22, 27 (left), 29 (top and bottom), 31 (top and middle), 36 (top), 40, 54, 70, 74, 84, 92 100-101, 117, 118, 130, 155 (top right), back cover (right)
Andy Jackson: p. 79
Anne Gallagher: p. 5, 104, 105 (centre left), 166
Bertie Gregory: p. 3, 58-59, 60 (top), 112-113, 115 (middle and bottom)
Blaise Bullimore: p. 138 (top), 159 (middle and bottom)
Drew Buckley: Front cover (bottom), p. 2, 15, 16, 17, 23, 32, 35 (top), 41 (top left), 43 (top), 103, 105 (top), 106-107, 108-109, 164, back cover (centre)
iStock: p. 50, 116, 120-121
Jake Davies: p. 69, 153
Janet Baxter: p. 33 (top and middle), 53
Jet Moore: p. 167
Kris Williams: p. 19 (top and middle)
Nick James: p. 4 (top left)
Dr. Rohan Holt: p. 4 (bottom), 167 (left)
Richard Kirby: p. 110-111
Rob Taylor: p. 167 (right)
Sally Weale: p. 5 (top), 166 (right)
Skomer MCZ staff: p. 71 (top right)

Just some of the crew who worked on *Wonders of the Celtic Deep*: Daf Matheson, Jack Carey, Matt Waddleton, Siân Phillips, Dale Templar and Olivia Tonkin.